Appraisal and repair of claddings and fixings

Titles in the *Appraisal and repair of building structures* series:

Appraisal and repair of building structures: introductory guide. Edited by R. Holland, B. E. Montgomery-Smith and J. F. A. Moore

Appraisal and repair of claddings and fixings. M. Wilson and P. Harrison

Appraisal and repair of reinforced concrete. R. Browne and D. Pocock

Appraisal and repair of steel and iron. J. Sutherland

Appraisal and repair of foundations. R. Fernie and H. St John

Appraisal and repair of timber. P. Ross

Appraisal and repair of masonry. G. Parkinson and J. Beck

Appraisal and repair of building structures

Appraisal and repair of claddings and fixings

M. Wilson and P. Harrison

Thomas Telford, London

Published by Thomas Telford Services Ltd, Thomas Telford House,
1 Heron Quay, London E14 4JD

Distributors for Thomas Telford books are
USA: American Society of Civil Engineers, Publications Sales Department,
345 East 47th Street, New York, NY 10017-2398
Japan: Maruzen Co Ltd, Book Department, 3-10 Nihonbashi 2-Chome,
Chuo-ku, Tokyo 103
Australia: DA Books and Journals, 11 Station Street, Mitcham 2131,
Victoria

First published 1993

A catalogue record for this book is available from the British Library

Classification
Availability: Unrestricted
Content: Guidance based on best current practice and research
Status: Refereed
User: Structural engineers

ISBN: 0 7277 1662 X

Typeset in Great Britain by MHL Typesetting Ltd, Coventry

Printed in Great Britain by Redwood Books, Trowbridge, Wiltshire

Contents

1

Introduction

This guide is one of a series which covers the whole field of the appraisal and repair of building structures. To reduce repetition the *Introductory guide* deals with all subject matter thought to be required for the generality of structural appraisals and gives overall guidance on repair. To assist cross-reference, headings for the initial chapters of this guide are identical to those used in the *Introductory guide*.

Cladding of buildings has been developed over the centuries. Many early Roman buildings were constructed using a thin facing material backed by brick or rubble walls to create a solid construction.

Building techniques and materials have improved over the years and fixing methods have become very important in order to attach adequately the facing material to the backing structure, which incorporates a cavity. Although actual fixing methods have not vastly altered over the years, the types of metal now used and the development of more sophisticated fixing elements have led to greater security. However, as natural facing stones in particular have become thinner, the boundaries of safety have, in some cases, been transgressed, and the combined problems of cladding and fixing elements have become more common.

With the development of precast concrete units as both load-bearing and non-load-bearing elements, the requirement for strong fixing elements has become more important, but it is often confronted by a lack of understanding of the complex movements to which buildings are subject. In the case of brickwork the problems of moisture penetrating into and across cavities has become a modern source of worry. Aggressive mortars have caused rapid corrosion of carbon steel and zinc-coated ties, to add to and combine with the moisture problem. Modern metal cladding has also given much cause for concern, although in many cases the problems have been caused by a poor initial design concept rather than a developing deterioration of a material.

It must not be imagined, or course, that all clad buildings are a potential source of trouble. Correct consideration given to all the aspects of cladding design will lead to a perfectly sound and adequate building. However, in all serious failures there are recurring elements of error which can be identified. The failures are often caused by initial design errors, lack of understanding of the possible construction difficulties and poor workmanship. The latter is often a cause on its own, but is usually also combined with the problems caused by the former two, i.e. design failure and poor understanding, not only of the design concept, but also of the reasons for construction requirements.

When investigating any form of failure, the main causes of the cladding problems must be understood, but at the same time an open mind is essential. It is all too common in modern contractual situations to look for a scapegoat rather than to identify a reason. It is therefore advisable to have a completely independent assessment. It must also be remembered that the solving of a problem is not achieved by merely replacing a broken stone, for example, by a new element. It is quite common for serious failure problems to be identified a second time around, simply because an inadequate making-good exercise was carried out in the first instance.

It must be remembered that the ultimate aim of any cladding repair programme is to return the building as closely as possible to its original design. In many cases this will be difficult if there are conservation orders on the property, or structural requirements which entail major structural repair or additional support to problem areas.

This book sets out very clearly the technical problems which can and do occur, as well as the solutions, and finally the analysis and design of the ideal cladding system.

2

Initial appraisal

The initial inspection is critical in determining the procedure to be adopted later. It is not sufficient to inspect a building visually and determine a solution without further physical inspection and material testing. However, a visual assessment will indicate the extent of the damage to the cladding material or its fixings, and from that initial stage a sequence for physical inspection may be determined.

The initial survey will need to be documented as fully as possible, and should set out the damage pattern. If possible a sequence of events leading to a culmination point (at which it was decided action needed to be taken) should be attempted. External inspection should be backed up by an internal inspection of the building; any internal distress will often give a good clue to the cause of external movement. However, too much reliance cannot be placed on such interior examination. False ceilings, partition walls, false floors and repainting can hide structural faults; it is essential to keep an open mind.

Photographs are also essential in any inspection. They are not only an immediate record, but may also be used to monitor continuing or future movement. It is obviously best to carry out such surveys in good weather conditions if at all possible; minute inspection of the surface of any material is not easily accomplished in wet conditions.

In tall buildings it will be essential to ensure that access is amply and safely provided by scaffolding. Mobile cranes (cherry pickers) and inspection cradles are often of considerable use and may even be essential. Abseiling has been used. Even in ideal conditions of general access, some areas may be inaccessible and it is advisable to include binoculars in the survey kit.

Factors to be considered in any survey will include environmental aspects, prevailing weather conditions, pollution factors and compass orientation. The past history of the building should be researched, especially any maintenance problems which were recorded. A building manager or maintenance supervisor are often good sources of information, and surprising facts can frequently be elicited from such people.

After all the damage statistics have been recorded, a fairly clear indication may usually be obtained of the possible causes. The appraiser should avoid setting his mind on a single cause, unless the problem is clearly defined, such as a broken fixing or perhaps (more rarely) a clear impact, a damage problem or rusting structural steel.

An assessment can then be made as to the next sequence of physical investigation. The requirement to open up the fabric of the building can be determined and any material testing may be specified. Often initial testing will be required, especially if certain aspects of the fixings are in question, such as bolt failures or a lack of fixings.

At this stage it is imperative that a concise report be prepared, setting out for the client the extent of the damage, the possible causes (if known) and future action. The total cost of repair work will not be known at this stage and it would be dangerous to hazard a guess. The client must, however, be advised of the cost of site investigation work and the implications that this might have on the continuing use of the premises. If the damage is serious and potentially dangerous, steps should be taken to safeguard the public against injury.

The second stage is to carry out the physical testing. Concrete cores may need to be taken, chemical tests may need to be carried out, and concrete casing to steel work may need to be broken back. A covermeter is a useful item of equipment. Stones may need to be taken from the walls, precast units removed or metal panels dismantled. If the damage is due to movement of any kind, reasons must be determined including, if necessary, any foundation investigation.

When all the results have been gathered a further report should be prepared and a remedial plan devised. Bills of quantity must be prepared and comprehensive quotations obtained. In any form of remedial work, however good the initial survey and testing may appear there will invariably be hidden problems, especially if the building is old, and due allowance should be made for such discoveries.

Once a comprehensive cost and remedial programme has been achieved, a clear instruction must be obtained from the client. Repair costs may be (and usually are) quite high, and often out of proportion to the value of the building. However, if it is considered advisable to proceed with remedial work, this should be done as well as possible, with no short-cut methods or half measures; such actions would lead to future problems, and in the long run would be uneconomical.

3

Signs of distress

The object of the visual inspection is to record the damage and diagnose the cause; this cause is not the same as the reason for the damage. Cracking in a unit may be caused by movement of the structure — what was the reason for the structure moving? This is the most important factor in any investigation. Repair related to cause will not cure the problem; one must diagnose the complete problem.

All forms of cladding are basically an overcoat to a building, and are therefore subject to the same problems. Generally, the effects are visually similar in most materials.

The basic reasons for cladding failures are as follows

- external
  - weathering problems (durability)
  - temperature effects (movement)
  - damage (impact, explosion)

- internal (behind façade)
  - lack of fixings
  - poor fixing (material or type failure, poor workmanship, etc.)
  - building structural movement (deflection, elastic shortening, foundations, etc.)
  - material breakdown (general attack, all types of corrosion, etc.).

All these reasons can be further split and the permutations are numerous; they are discussed in section 6.1. The movement effects are similar in various cases, and the usual signs of damage are as follows.

3.1. External

Weathering problems

All stone will weather and it may also be affected by pollutants. The wrong choice of stone or stone cut on the wrong bed will result in deterioration; surface damage will occur. Concrete will weather badly but will not necessarily exhibit damage unless porosity causes corrosion. Serious deterioration could be due to carbonation, or chloride or sulphate attack, requiring laboratory testing.

Brickwork may become porous and may allow moisture damage to occur, which will affect ties. Joint cracking will be obvious; slip brick will become detached. Metal facings will suffer in the same way when surface protection fails.

Timber facings are obviously subject to severe attack, and any material degradation should be readily identifiable.

Temperature effects

Movement is quite frequently the cause of cladding problems. In concrete, stone and brickwork cracking will occur as an effect of movement, often following bedding and vertical joints; however, if the various elements are well retained and fixed, the cracks may also pass through the various units. Heavy parapet coping stones will move bodily under temperature stresses. In the height of the building, if compression joints have not been provided the wall will expand and cause crushing at support levels, giving rise to spalling of the brick, stone or concrete.

Slip bricks will be displaced by a change of temperature. In the case of concrete units, which are normally substantially fixed, quite serious local damage will occur around fixing points. The effects in these cases can be quite serious, including the deformation of fixing brackets and bolt failure.

Metal cladding will be distorted, and if allowance has not been made at connection points, tearing can occur. Glass-reinforced cement (GRC) is particularly susceptible to temperature damage, particularly so if the units are large, shaped and dark in colour. Cracking will be obvious, and again damage can occur locally at fixing points.

Impact damage
Unforeseen damage due to external events will occur, and this will be obvious. However, there are events which may be anticipated and should be allowed for in design. Unfortunately, all too often this is not done. The type of impact damage which occurs most frequently in stonework clad buildings is caused by ladders propped against the wall and window cleaning cradles, or similar equipment. A very frequent point of damage is the base course of stones on a building.

Cavities at these levels should be adequately backed so that the stones cannot be broken by an impact. An area where this type of damage frequently occurs is around lift shafts and stairs; damage of this type can generally be easily seen.

3.2. *Internal*

Poor fixing
The damage due to fixing problems is usually obvious externally; elements will become misplaced. This type of failure is among the most serious, and consideration should be given to any inherent risk to which the situation may give rise. Brickwork with badly installed or corroded ties will be unstable, and stability must be of prime importance.

Metal cladding is often relatively lightly fixed to backing sections, usually light extrusions. Self-tapping screws are the most usual method; these may be torn out or may even suffer bimetallic corrosion.

Building movement
This event is an area of great trouble, and can of course affect the whole building. Its effect on the cladding is generally similar to that of straightforward temperature movement; however, many more factors are involved. One should look for settlement, unevenness in doors and windows, and movement over openings

(possible structural deflection). There will be displacement, rotation, cracking and crushing when applied to stone, brickwork, concrete and GRC. Metal will crease, buckle or ripple.

Material breakdown
The most obvious manifestation of damage will be environmental; however, in the context of the building as a whole, structural deterioration of the main building can affect the cladding in many ways which are not immediately apparent.

Breakdown of the background structure can cause loosening of fittings. Corroding steelwork will push forward brick or stone claddings, and precast concrete will move. These are not easy problems to identify, and only investigative survey work will identify these situations.

4

Signs of maintenance and alteration

As noted above, failures requiring drastic repair are often second-time-around problems. It is therefore important to look for signs of previous repair; these will indicate if the current problem is related to any previous repair work.

Continuing spalling of concrete surfaces is a firm indication that there is a chemical or corrosion problem. Reopening of cracks in brickwork or stone will indicate continuing movement trouble or the corrosion of structural steel.

The opening up of an existing façade may reveal many disturbing things. A poor backing wall comprising bad brickwork or poor concrete can be the cause of loose or even missing fixings. Perhaps the original fixing elements may not have been to a required standard.

It is, of course, difficult in many cases to determine the original specification, and one can only be guided by present standards or perhaps knowledge of past methods. The omission of bolts is not uncommon and the inadequate tightening of bolts is also a factor to be noted. In the investigation of previous repairs, alteration to fixings, signs of removal of the original fixings and redrilling of positional holes are all points to be looked out for.

A common type of brickwork failure, in addition to brick tie corrosion, is slip brick detachment. In some cases previous loose tiles may have been repositioned on fresh mortar beds without the use of new mechanical fixings.

The causes of failure of this nature are numerous: poor or non-existent compression joints, poor mortar bedding, the lack of a cavity tray, bad alignment of the concrete backing nib, and uneven thickness of slip brick tiles where purposely cut to suit site dimensions. All these signs must be looked for.

Poor repair of previous cracking in cladding elements by the use of incorrect or incompatible materials may be seen. It has been known for bolt holes to be drilled and the expansion anchor not inserted, but rather a bolt or set screw has been hammered into the hole.

In older buildings where failures have manifested themselves some years after construction, 'repairs' may well have been carried out by jobbing builders or maintenance men. Such repairs will often be obviously amateurish, and have not taken account of the initial cause.

There are common factors in many cladding failures, but nevertheless assumptions based on poor knowledge should be avoided. Full study of all the facts resulting from any investigation should be open minded at the same time, backed by prior knowledge.

In cases of metal cladding panels may have been replaced without correctly refixed sealing gaskets. This will, of course, also apply to any sealed face elements; signs of new inserted sealants are points to look for. In the case of rendered finishes, one should be careful to ensure that newly covered areas are not cosmetic, hiding previous cracking (in brickwork primarily) which has not been repaired.

5

Methods of investigation

The initial survey is generally termed 'visual'. This does not of course, rule out some basic probing. Severe damage should not be caused unless on-site permission has been obtained during the course of the inspection. It is always an advantage to carry out early inspections with the client or his representative in attendance.

Mortar joints can be checked, timbers can be probed, mastic can be checked for plasticity. Small sections of concrete spalled edges can be broken off and pieces of stone taken out, if there is not too much disruption. Walls can be tapped for hollowness behind renders or faced stonework. In addition, glazing putty may be checked and flashings and DPCs, etc., inspected. These are all items which can become affected or themselves cause problems.

One should not, however, open up sections of the façade unless they can be made temporarily weatherproof. It should be remembered that all cladding repair contracts are fairly long term, and protection of the building does become an item of major importance. Scaffolding costs are high, and adequate temporary sheeting, and perhaps temporary roof staging and protection, will be needed; it is these items which add tremendously to the cost of repairs.

It should also be remembered that if replacement units such as precast panels, new stonework or metal panelling are required, these may have extended delivery periods. Protection is therefore of prime importance.

More thorough investigative and follow-up surveys will require access to be available. As previously noted, safety must be a prime consideration. Full scaffolding will not usually be available at this stage, as the extent of repairs will depend on the final decisions and/or contract programme stages.

Listed below are the types of investigation which might normally be required.

- settlement problems
  - boreholes
  - trial pits
  - soil tests (laboratory)
  - site
  - drainage tests
- movement
  - Vernier
  - graduated telltales
- material samples
  - concrete cores: laboratory testing for sulphates, chloride and alkali reaction, and carbonation, strength, material type, etc.
  - stone samples: strength, absorption, etc.
  - metallurgical tests

Generally there should not be a material problem regarding steelwork.

Regarding fixings, although stainless steel is now generally advised and specified, copper alloys were used for some years before the 1960s, and some of the bronzes, particularly manganese bronze, high tensile brass and copper zinc alloy, have been found to be subject to stress corrosion. Such fixing materials, if located, should be tested for their chemical composition and replaced in the works.

The location of fixing can be carried out externally with the aid of metal detectors. A fairly sophisticated instrument is advised, one that differentiates between the reactions of various metals. Austenitic stainless steel is non-magnetic, and therefore will not be easily located on some instruments.

Cavity inspection may be made using optical fibres, preferably with a camera attachment. For quick in situ testing of concrete in order to give a reasonable guide to strength, a Schmidt hammer may be used. It must be remembered that laboratory testing is relatively expensive and time-consuming, and due allowance must be made in the budgeting, not only for cost but also for pre-contract time.

6

Types of failure

6.1. Cladding failures and remedial works

What constitutes a failure? It should be remembered that a cladding failure is not necessarily a total collapse causing damage or loss of life. However, the failure of a cladding fixing can lead to instability in the façade and it constitutes a potential hazard.

Considering the numerous types of cladding material and the various forms of structural construction material, there are many combinations of potential failure, but all are dependent on basic factors. It is probably sufficient to define four areas

- structural building failure
- cladding failure
- failure of fixing methods/fixing components
- workmanship.

It is often a simple matter to determine the type of cladding failure and to propose a remedy, but a quick solution is not always the correct or economical answer. It is generally recognized that most serious cladding failures are identified a second time around. The problems therefore go deeper, and it should be determined why the failure occurred.

Structural failure

The total structural failure of a building in this country is rare and generally the lessons learned from recent ill-conceived system building collapses have been heeded. Isolated member failures, however, are not uncommon and the most readily identifiable kinds are due to excessive deflection movements on beams and floors, both in concrete and steelwork. Causes of such failures may be complex, e.g.

- poor concrete
- poor workmanship
- combination of both of the above
- physical overloading
- bad design
- omission of reinforcement
- chemical attack and corrosion.

In the case of serious building settlement there are factors which are unknown or unavoidable, or are simply basic errors. The list is endless, but basic consideration should be given to

- poor soil conditions
- lack of site investigation
- incorrect choice of foundation
- footing failure
- pile failure
- design error
- undermining of existing foundations
- construction of tunnels, etc.
- mining subsidence
- ground slip
- differential settlement within or between buildings.

Building movements other than settlement causes can be due to

- temperature movement with excessive shrinkage and expansion
- compression movement due to elastic shortening under load
- vibration caused by vehicular traffic, machinery or underground railways.

These types of structural failure can occur in all the major forms of construction.

When a cladding 'failure' has been identified, all the above possibilities must be investigated to determine the basic cause of the cladding movement. Sometimes, of course, the cause of cladding failure may be obvious, such as insufficient fixings, but nevertheless why did the movement manifest itself in the first place?

Cladding failure

There are many reasons for cladding failure, but again basic areas may be identified

- structural failure
- material faults
- fixing faults
- lack of movement joints
- workmanship.

The causes of structural failure noted above may be apparent in the cladding as excessive movement of panels or individual stones or groups of stone (Fig. 1). Deflection causes excessive compressive loads to be transferred to the cladding, causing crushing. This may be manifested by spalling, cracking, deflection, or misalignment, depending on the cladding material. Compression movement causes similar results.

Expansion and contraction can cause the cladding to crack. Often it can be clearly seen that the cladding is moving more than the main structure, as may be readily understood since the outer layer of material will always be more affected by thermal movement, particularly when cavity insulation is present. This is why regular or sensible placing of movement joints should always be considered in the initial design. However, investigation may reveal the causes to be induced by the creep of the main structure; this may cause isolated points of excessive deflection and failure in the structure and the cladding.

Material failure

Material failures can be of two types: natural and inflicted. Natural faults will include geological imperfection in the material, in the case of natural stone; unnatural or inflicted faults will include

- stones not cut on the appropriate bed for their application
- in the case of precast concrete, material imperfections, poor compaction or lack of cover to reinforcement
- design and installation faults causing moisture penetration, freezing and subsequent spalling; debris in movement joints

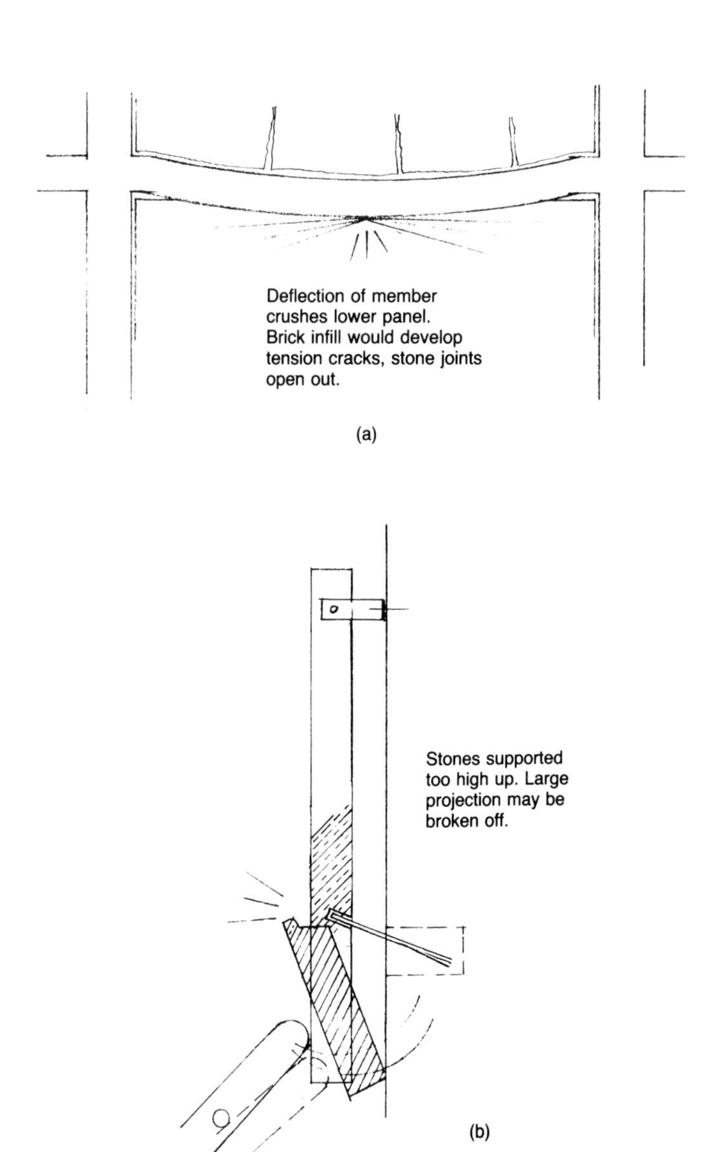

Fig. 1. (a) Deflection; (b) accidental external damage

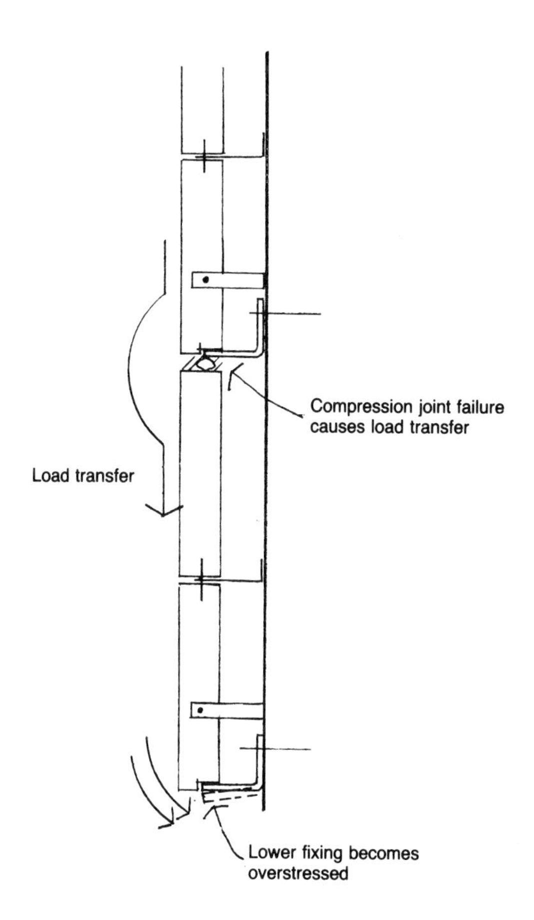

Fig. 2. Joint failure

causing localized stress under movement (Fig. 2)
- design faults: stone area too large, causing stones to fail under cyclic wind load or applied equipment loads; insufficient joints in material, disregard of movement joint or thermal bow.

Fixing faults

A fixing failure is rare, although in the case of heavy precast panels distortion may take place or the fixing may be pulled from the parent material. Any complete failure could be due to

- incorrect choice of fixing metal
- use of dissimilar metals
- incorrect choice of fixing
- poor installation
- failure due to overstress
- metal failure through a manufacturing fault (rare).

Workmanship

This is the most contentious area, and in nearly all cases of failure workmanship would become the scapegoat. Apart from the very clear areas of blatant poor workmanship or disregard of instructions, the reasons can be much deeper

- poor installation
- omission of fixing, intentionally or accidentally
- abuse of fixings, overtightening bolts, hammering bolts through or into undersized holes, site 'adjustment' to take up dimensional discrepancies
- site alteration to fixing locations in the units or stones
- site abuse by theft.

These are some of the reasons of failures, but one must consider the causes

- poor or inadequate instructions
- unfamiliarity with the fixing method
- poor site supervision
- poor design concept allied to poor instructions and possible subsequent omission of fixing.

Omission is often, in brickwork failures, a deliberate case of omission due to speed and irresponsibility. However, one must consider the case where a fixing cannot be easily installed — it is left out. Irresponsible, yes, but whose fault is it? Is it poor design, poor choice of fixing or poor supervision? The omission of a fixing is not always done irresponsibly, it can be done in ignorance; there is no excuse in law, but who is responsible?

Often a fixing cannot be installed because post-drilled holes foul reinforcement or steelwork. A cast-in fixing may be misplaced and the fixing attachment is misaligned. The hole is either moved or ignored; if it is moved it can lead to 'alteration' in the fixing to suit the new position and alteration in the location point in the cladding element. This action is well intentioned, but without the designer's agreement it is possibly disastrous.

The last fixing in a run of stones may be omitted owing to the difficulty in locating it. Site dimensional errors are also a source of danger where fixings are 'adjusted'.

Corrosion can also be a major factor in fixing failure. The presence of water is of prime concern, and therefore correct detailing of cladding elements is vital to prevent the ingress of moisture at joints and connections. After erection the fixings are inaccessible, and cannot therefore be checked without dismantling the units.

Another source of trouble is overtorqued bolts. Many erectors consider that they can adequately 'tighten'a bolt, but tightening is not the same as torqueing. A torque spanner must be used. Invariably, any bolt up to and including M12 dia. will be overtorqued, and any size above will be undertorqued, if it is tightened without the correct torque spanner. It is also essential that torque spanners are regularly checked and correctly calibrated.

This chapter has so far concentrated on mechanical fixings; however, there is another important field, which includes glues, resins and mortars. Common failures with decorative facings such as mosaic and slates may be due to incorrect or badly applied adhesives.

Resins are generally sound. However, age and correct storage conditions are essential if the resins are to be effective when applied. Resins and pumped mortars should be avoided in porous materials, and especially in old and suspect brickwork. Two-part mortars are often a problem, especially in adverse weather conditions, and mixing instructions must be very strictly adhered to.

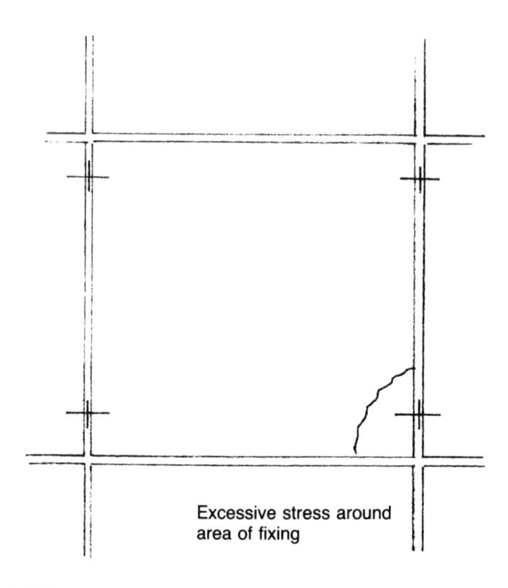

Fig. 3. Local failure

Some situations may call for the mortar components to be kept in a refrigerator on site to avoid rapid deterioration in the heat.

The above sets out the basic causes of cladding failure. A study of the cases will reveal certain common facts which stand out clearly in every investigation. There are basically four prime causes

- lack of awareness
- lack of supervision
- design inadequacy
- workmanship.

6.2. Modes of failure

There is a very small number of ways in which cladding can fail. These ways may be identified initially by visual inspection before any expensive opening up is carried out, although experience will soon enable an assessment of the potential seriousness of the situation to be made.

A reasonable check list is as follows.

Complete detachment from building (Fig. 3)
Causes: unit spalled away around area of fixing, fixing collapsed, or no fixing at all.

Cracking of unit at fixing point (Fig. 4)
Causes: movement of fixing, too little backing material to fixing to resist load, or structural movement causing misalignment.

Cracking of unit within body of element
Causes: undue stress applied to unit through fixing points, i.e. temperature movement, structural movement causing misalignment, or accidental loads from external sources.

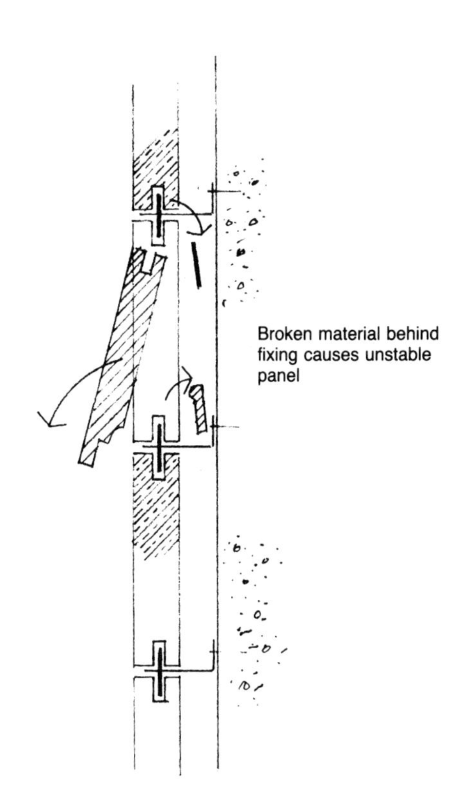
Broken material behind
fixing causes unstable
panel

Fig. 4. Material failure

Spalling of unit
Causes: if natural stone there could be compression loads due to structural movement or temperature movement, compression loads due to joint failure, corrosion of reinforcement in precast concrete, frost action through freezing of trapped moisture in crevices and joints, general chemical attack, or corrosion of fixing elements.

Misalignment
Causes: building movement; fixing movement; instability for various reasons; expansion, contraction forces, wind suction and pressure (not unusual in large areas of brick panel cladding moving on DPC).

These modes of failure may be clearly identified visually, and the accompanying figures indicate the various results.

The problem remains of what to do.

7

Remedies for cladding failures

7.1. Brickwork

The most familiar type of brickwork failure is that due to the corrosion or the omission of wall ties. There are many remedial ties available which are technically simple, and these include metal straps and rods, which are either mechanically activated or bonded using resin or grout. Some remedial systems require bricks to be taken out before fixing into the inner skin and later making good. Although this is technically simple, some degree of skill and experience is essential in order to ensure success, bearing in mind that brick ties rely on the compression force from above to achieve their working figures.

An occasional fault in brickwork may be identified through the use of Pistol bricks. These are usually positioned on continuous angle supports, above compression joints. Unless joint filler is placed in the joint before the bricks are bedded onto the angle, the mortar may squeeze out into the joint and the brick will rotate. This gives initially poor bedding for the upper brickwork; it also deposits mortar into the compression joint, rendering the gap ineffective as a joint unless raked out.

Occasionally it will be found that the inner skin of a cavity wall is too poor to accept a remedial fixing, especially a mechanical expander. In such cases it is often sensible to demolish and rebuild the bad sections of the wall.

When the failure has been caused by corrosion of strip ties,

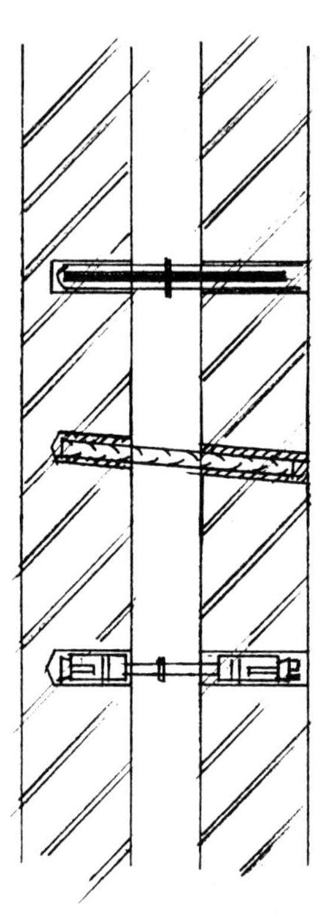

Fig. 5. Types of cavity wall remedial tie. BR Digest 329 refers to various remedial ties. Three effective methods are a bar in resin grout, reinforcement in an inner resin capsule, and an injected grout outer skin. The bottom drawing shows a mechanical double-ended expander

which often starts in the outer leaf, every effort must be made to remove all trace of the offending tie, as failure to do so will mean that the wall will continue to crack along its bed joint due to the expansion force of the corroding tie. Any decision to remove the offending part of the tie from the inner leaf must be based on the evidence of damage through corrosion, compared with the possible internal damage to decoration.

Other causes of a wall tie failure could be damp penetration caused by porous brickwork, or poor or chemically aggressive mortar.

General brick wall failure may be due to settlement of one form or another, and in such cases, provided the cause is correctly diagnosed, rectification of the structure will be more significant than any isolated remedial work to the brick wall. It is therefore, as previously noted, essential to diagnose the cause correctly, not just the isolated problem.

Large panels of infill brickwork often fail due to a lack of provision of compression joints at the top and sides; this results in crushing, cracking and misalignment. The usual solution is to cut out the brickwork and provide sliding ties or dowels fixed to the structure, and to introduce a flexible construction joint against the structural member.

In long runs of brickwork joints can be saw-cut into the brick wall at intervals (Fig. 6). Provided the wall is generally stable this will be satisfactory. Any unstable sections should, of course, be replaced or at least stabilized satisfactorily.

Cutting joints into existing brickwork will leave free ends, and the standard spacing of wall ties, if the unit is of cavity construction, will need to be augmented with additional remedial ties of some kind.

One of the most common forms of brick facing failure takes place on parapet walls, and is especially common if the backing member is formed as a concrete upstand. In these cases the brickwork should be taken back about 1·5 m from the corner and rebuilt with vertical movement joints on each face. Usually it will be found that the concrete upstand has also cracked, and, provided the member is not a spanning upstand beam, consideration must also be given to jointing this member.

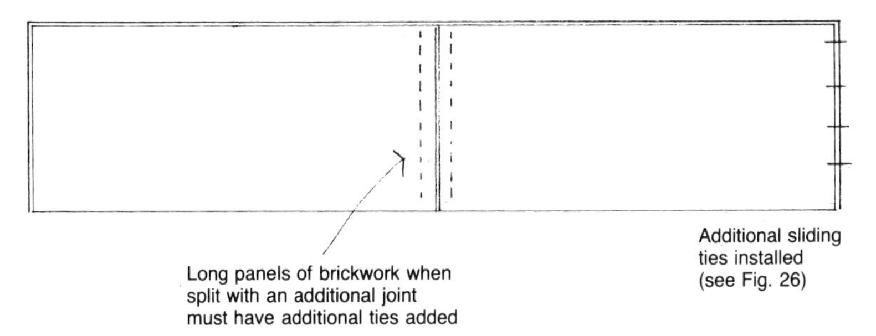
Long panels of brickwork when
split with an additional joint
must have additional ties added
Additional sliding
ties installed
(see Fig. 26)

Fig. 6. Stabilizing long panels

7.2. Remedy for concrete panel failure

The failure of concrete panels present very different problems, and, if due to movement, can have devastating effects on the fixings. Movement is one of the prime causes of facing panel failure, and often bad design of the fixings is the basic cause. Units bolted tightly together, or even in some extreme cases welded to the main structure, are common. Flexible joint separators should always be used between units. Such failures need individual consideration and removal; replacement will often be found essential.

Other forms of failure in concrete units include spalling and cracking due to reinforcement corrosion, caused either by chemical action, porosity or lack of cover. In such cases, repairs to the faces of the units can be effected satisfactorily. It is again essential to remove badly corroded reinforcing bars, or thoroughly clean and treat them, replacing bars as required. If chemical action is responsible, and it is identified as serious and progressive, replacement must be considered. These forms of unit failure do not normally affect the fixings, always accepting that the fixing was not the cause of the initial problem. Again, it is important to determine whether or not bimetallic corrosion is a cause of the problem.

7.3. Natural stone less than 50 mm thick

Slate, marble, travertine and granite are the materials that most commonly fail. These failures are generally attributable to movement or imposed accidental loads.

It is not a simple matter to repair stones in situ and often the failure will be localized. To take off a whole wall of stones for an isolated problem is costly. It may be possible to work back from an end stone, but generally the only sensible solution will be to face-fix the stones. It is possible that movement perhaps caused by thermal forces is due to a shortage of fixings or a lack of movement joints. This is a typical case of identifying the underlying cause and not the result of an action. The fixings can be located either visually by removing panels, if possible, subject to the constraints noted above, or by using metal detectors or optical fibres.

In this method it is essential to use a sophisticated instrument which identifies the difference in magnetic potential between various metals; if stainless steel fixings have been used their potential is different from that of mild steel. This was a method used in identifying some incorrect dowel fixings on a large contract where bright mild steel dowel pins had accidentally become mixed with stainless steel dowels. Visually they were indistinguishable; however, austenitic stainless steel is substantially non-magnetic.

Remedial face fixing also has its problems, however. There are two basic difficulties: not knowing what is behind the façade and the actual difficulty in drilling through the stones while they are actually on the wall. It is unlikely that a suitable drill bit could be used on a hard stone without shattering it unless diamond drilling is employed, but on no account should percussive drills be used. Examination of stones under laboratory conditions indicates that drilled stones do exhibit faults which might possibly cause later cracking. The seriousness of this is disputable, but nevertheless the evidence does indicate that care needs to be exercised in any form of drilling. Ideally, stones should be removed in order to drill for a face fixing, and if this is well done, with well-installed plugs carefully matched to the stone, they are difficult to detect. Evidence is available that some types of marble will bow when in use, particularly if the panels are 20–30 mm thick. Consideration should be given to the maximum panel size in relation to its thickness, and advice should be sought from the stone producer. Face fixings can also be used in original installation designs, and at the initial stage of building this is a very acceptable method. However, if all else fails, complete removal and refixing is the only answer.

Modern remedial work sometimes includes overcladding, but this should only be considered in cases of environmental degradation. If there are signs of building failure these must be examined and corrected initially, not just hidden; that would be merely a potentially expensive cosmetic repair.

7.4. Natural stone greater than 50 mm thick

Failure of heavy natural stones may usually be attributed to the omission of cramps, the non-location of dowels or a lack of movement joints. The latter is a prime cause; the remedy is to remove the stones and refix. Occasionally the stone may fail because it has been wrongly bedded, but this is rare.

7.5. *Brickwork/timber frame*

Although not specifically covered in this book, some general points can be made about this combination. The important thing to watch is the relative shrinkage movement between timber and brickwork; this movement can be very substantial, and ties must be capable of resisting the loads specified as well as the degree of movement expected.

7.6. *Cosmetic finishes*

Generally, experience with mosaic and tiles has not been good. Quite simply, anything stuck to a structure has a fair chance of failing. The adhesive may not be suitable or may have been wrongly applied. The subtrate needs to be completely clean, and this is often a major factor in failures. Whatever the reasons, one should avoid sticking cladding of any description on a building unless a reduced life expectancy is acceptable.

7.7. *Summary*

Cladding failures may be sudden and may take unexpected forms. The most important thing is that any large cladding contract should be considered at the design stage and should not be left to an inexperienced site-fixing contractor. It is essential that the methods and the sequence of construction be considered from the outset of the design. It will pay to seek early professional advice; repair costs can easily outstrip the initial cost of cladding. In addition, and very importantly, supervision must be strict and well controlled.

8

Forms of failure

8.1. Concrete panel

The failure illustrated in Fig. 7 is caused by movement of several types, and it clearly shows the very high forces which can be induced. There is evidence of temperature movement, elastic compression, out of balance rotation and faulty installation. Any one of these situations can be bad on its own; here all are combined.

In an analysis of the case, the following points may be noted

- rigid fixings
- obvious centre of gravity action outside line of support
- long panels inducing considerable movement
- channel cast into column and bonding straps not bent back (most are now supplied preformed)
- edge distance from channel to column corner, which reduces effective 'cone' area and accelerates early tensile failure
- cracked support nib.

The solution to this type of problem is complex and the most difficult exercise is actually to assess the true forces. This problem includes a full fixing failure.

Various tests have been carried out on test pieces to determine the ideal and indeed essential aspects of cast-in-channel fixings. Experiments with various configurations of strap are noted and the results are interesting.

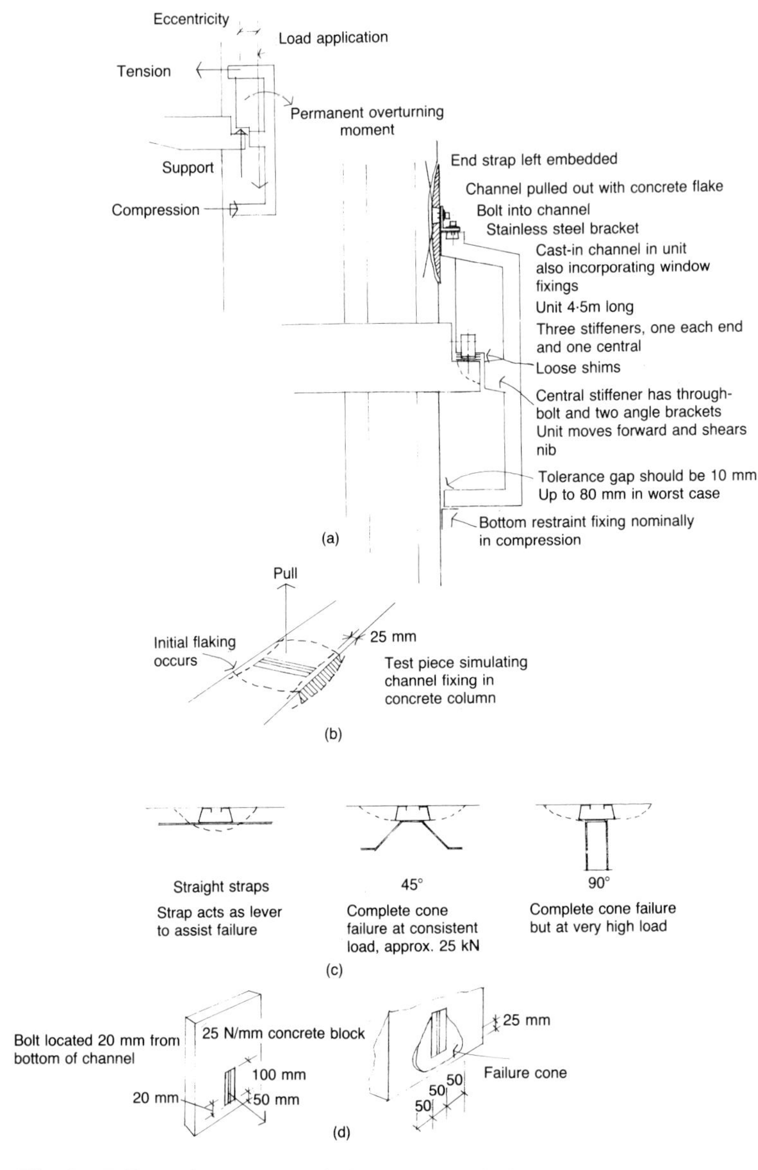

Fig. 7. Failure of concrete panel: (a) precast unit fixing failure; (b) channel failure mode; (c) cast-in channel restraint strap configuration; (d) test failure of eccentric channel

Straps have been tested straight, at 45° and at 90°. The results confirm the following points.

- The strength of the concrete makes little difference; however, the concrete must be of an acceptable structural standard, and for a channel used in lightweight concrete a reduction in the working load due to increased deflections is probable.
- The failure is always in the concrete, never in the bolt or channel, provided the loads are such that the bolt sizes are 12–16 mm in diameter.

What such tests show is that it is essential to ensure the adequate bonding of the straps

- straps should certainly be bent back (the angle seems immaterial, provided the strap is not straight)
- the straps should preferably be set behind reinforcement bars
- there should be an adequate distance between the centre of the channel and the concrete free edge.

Normally a resistance to pull-out would give 25–28 kN with nominally bent straps, say at 45°. With straight unbent straps the initial concrete failure is around 14 kN and the strap appears to act as a lever which self-perpetuates the failure. Vertical straps are obviously ideal and the induced concrete failure is in the region of 30 kN. (During the tests loads of 50 kN were tried and the tests were abandoned owing to potential failure of the testing frame.)

Other individual tests have revealed how important the edge distance can be, especially when the bolt is located eccentrically in the length of the channel.

8.2. Cast-in sockets

This example does not in fact apply to one specific problem, but it contains knowledge gleaned from several problems, all exhibiting a common failure mode.

Cast-in sockets are of two types: one specifically for lifting, the other for general fixing. Often the single socket performs a dual role, but it must be capable of taking the worst loading situation.

Lifting sockets do not normally fail under conditions of lifting unless some basic incorrect action is being applied. Lifting sockets should normally be linked to reinforcement, or have a bar through them which laps with or is tied to the panel reinforcement. However, failures do occur. The example (Fig. 8) is of a lifting socket placed at a change in section of a unit, where the unit is lifted on 'brothers' which are too short. Additionally the lifting point is not placed on line with the centre of gravity.

A similar example shows a shaped unit being incorrectly lifted, inducing such excessive stress that the sockets shear.

Neither of these types of failure would have occurred if the units had been correctly stiffened with a lifting frame or strong back of some kind. Shearing of sockets also occurs if the lifting bolts do not have sufficient thread engagement, and the point of failure is across the body of the socket at the base of the thread where the bolt does not penetrate.

8.3. Glass-reinforced cement failure

Figure 9 shows a typical 'failure' mode and it demonstrates all the classic elements of GRC. Although in itself an admirable material, it was probably used too ambitiously and it is now recognized that its behaviour is not constant with age.

The evidence does indicate that, however thorough the pre-launch testing of a new material may appear to be, its use in service cannot be fully predicted. This particular example exhibits the major factors now identified as contributing to the failure of GRC sandwich panels

- size
- shape
- colour
- location.

Investigations on other projects indicate that other worrying factors also contribute, poor workmanship and bad installation being foremost. It must be remembered, of course, that this, although in terms of a cladding unit a failure, is not a serious danger to life and limb. Movements of this nature do not necessarily lead to total collapse.

A unit may be repaired and fixings replaced, but one must also consider the long-term economics. If an industrial building is being considered, what is its expected or planned life, and will the ongoing deterioration take so long as to make an immediate and

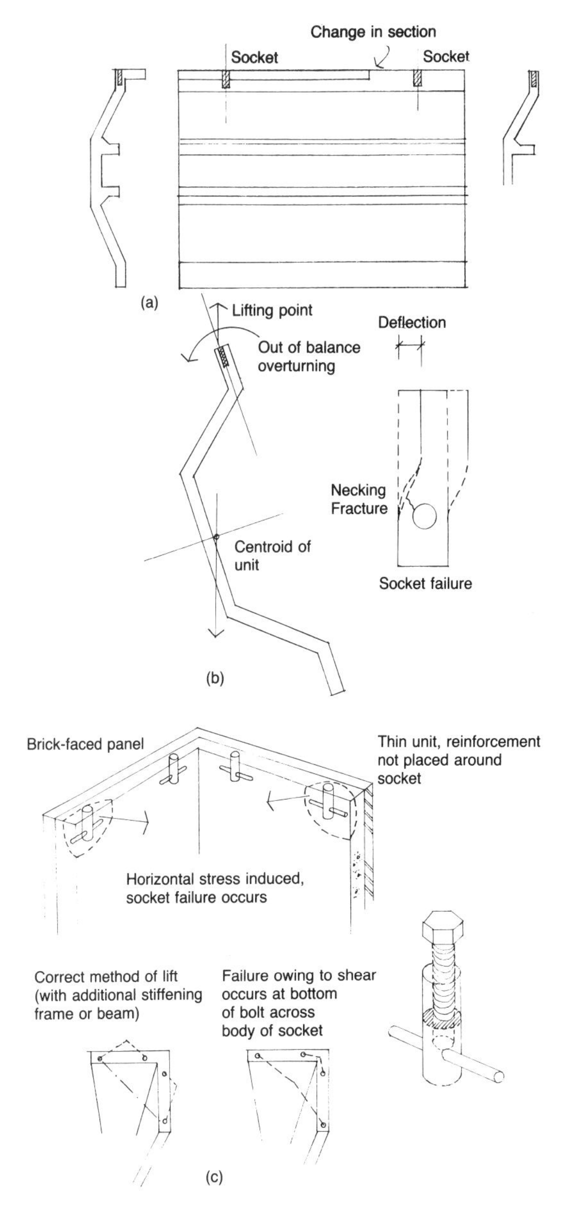

Fig. 8. Failure of cast socket: (a) concrete unit; (b) unit lifting problems; (c) method of lift causing failure

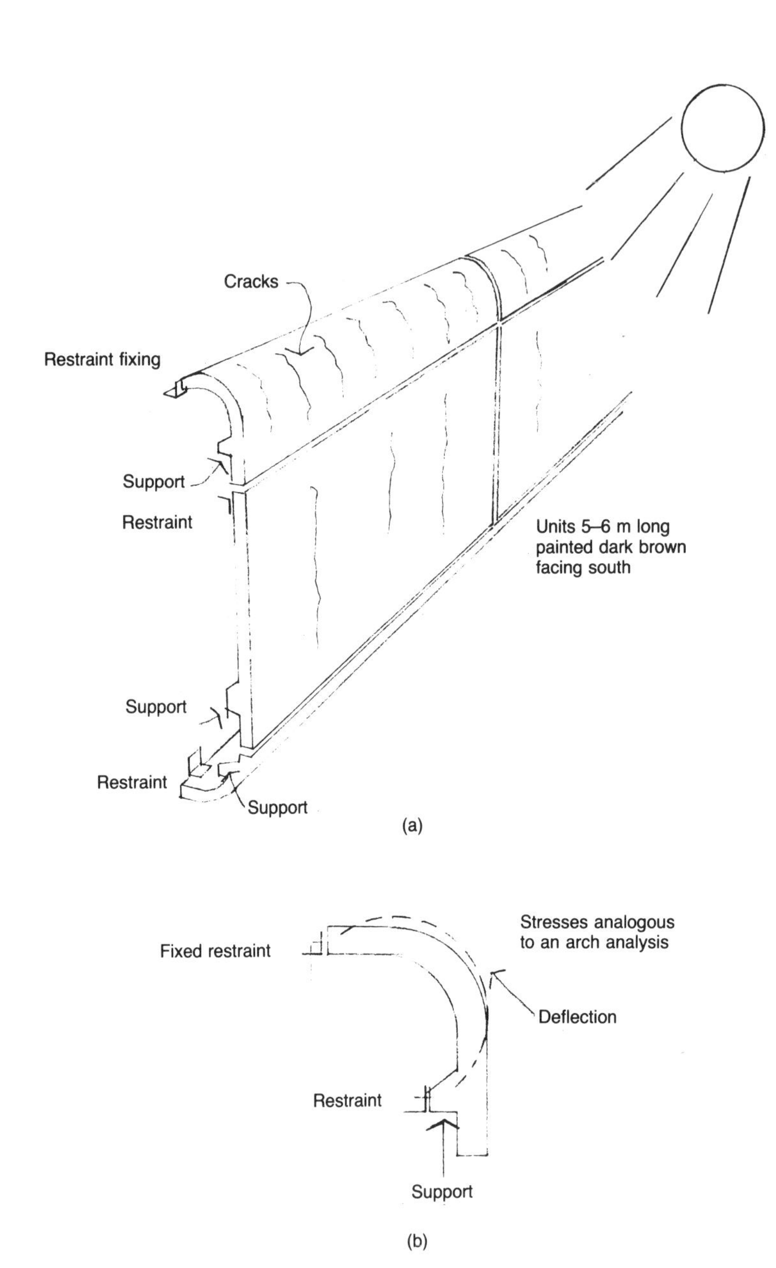

Fig. 9. Failure of glass-reinforced cement: (a) typical failure pattern; (b) failure mode

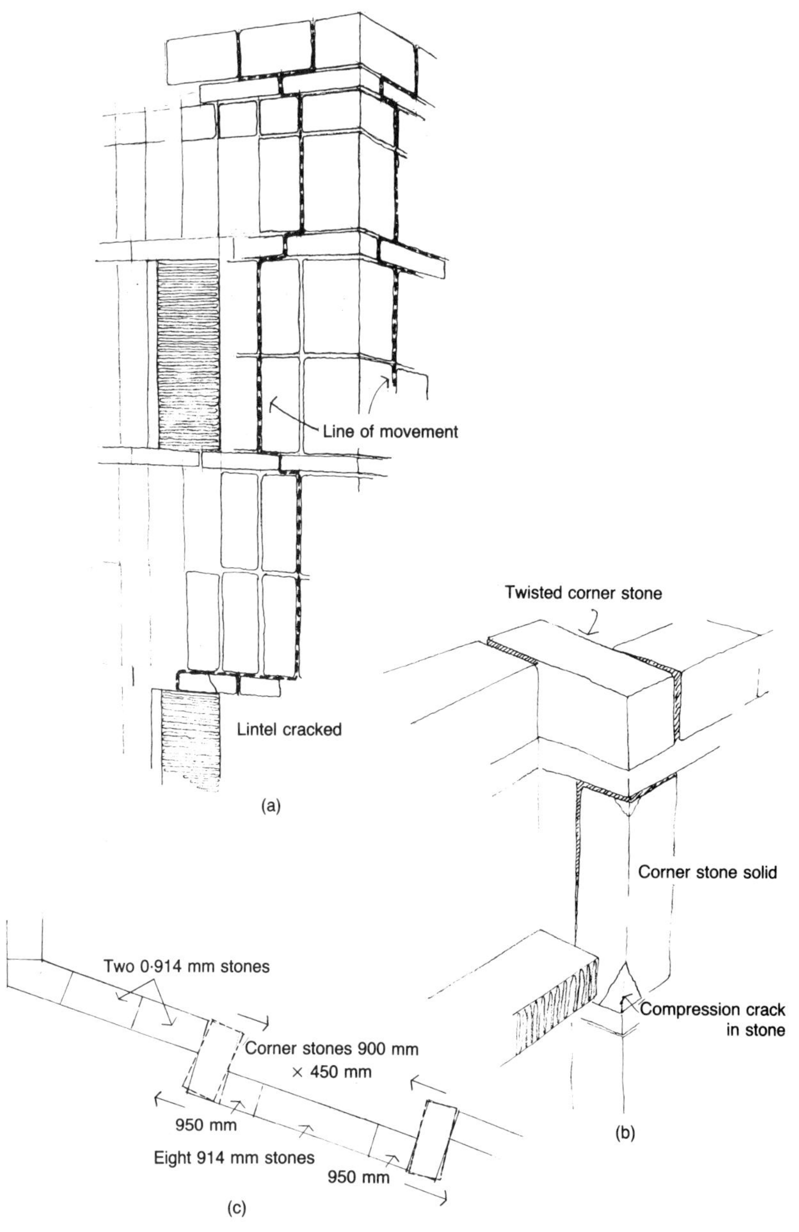

Fig. 10. Failure of natural stone: (a) movement of heavy stone façade; (b) corner stones; (c) plan of parapet stones

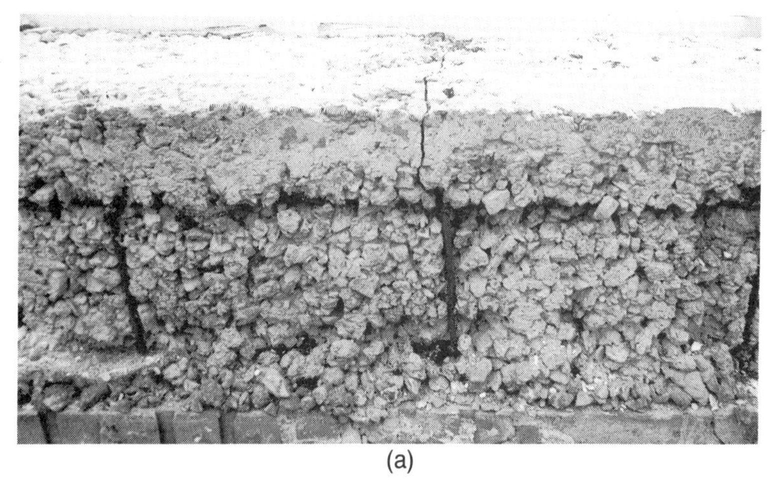

(a)

Fig. 11. (a) Parapet beam after removal of facing and coping stone; (b) general condition of wall faces throughout building after removal of stone slabs; (c) view showing coping movement and shift of upper stones

expensive repair or replacement viable? All these points need to be considered in assessing the extent of remedial work.

8.4. Natural stone

The example shown in Fig. 10 is part of a case history, and it indicates the type of movement which takes place when thermal forces are high. Typical of many cladding problems, the parapet movements have a progressive effect on the facing stones.

One important point to note is the size of the parapet stones, and when the weight is considered it may be seen how high thermal forces can be.

The repairs on a contract of this nature are fairly straightforward. The stones may be face-fixed and the parapet's existing joints may be refilled (the existing mastics have hardened) and new ones introduced. At the ends and corners the parapet stones should be lifted and new slip planes introduced over the heads of the wall facing stones, allowing future independent movement. This condition is fairly common, although in some cases it will be found that the causes are various, including missing or inadequate fixings.

(b)

(c)

The following examples shown in Figs 11–14 are not identified as specific contracts, and are noted as typical situations, combining two or three actual examples.

(a)

(b)

Fig. 12. (a) Corner with exposed frame showing severe deterioration of reinforcement causing stones to be forced away from building; (b) reinforcement was cut away and new bars bonded in and repaired; frame was temporarily supported with stainless steel plates, as repairs could only be achieved from outside building; (c) general condition of backing wall before renovation; a complete stone unit can be seen built into backing wall; (d) wall cleaned off and prepared for refixing; (e) stones being reset with new restraint fixings in stainless steel; original ties were of manganese bronze many of which had snapped due to embrittlement

(c)

(d)

(e)

(a)

Fig. 13. (a) Slip brick replacement to concrete support nib; wall panel supported on temporary stools during removal of upper courses of brickwork; (b) replacement of cavity tray, additional ties and slip brick ties in position; (c) rebuilt brickwork and nib prepared to receive ties mechanically retained

(b)

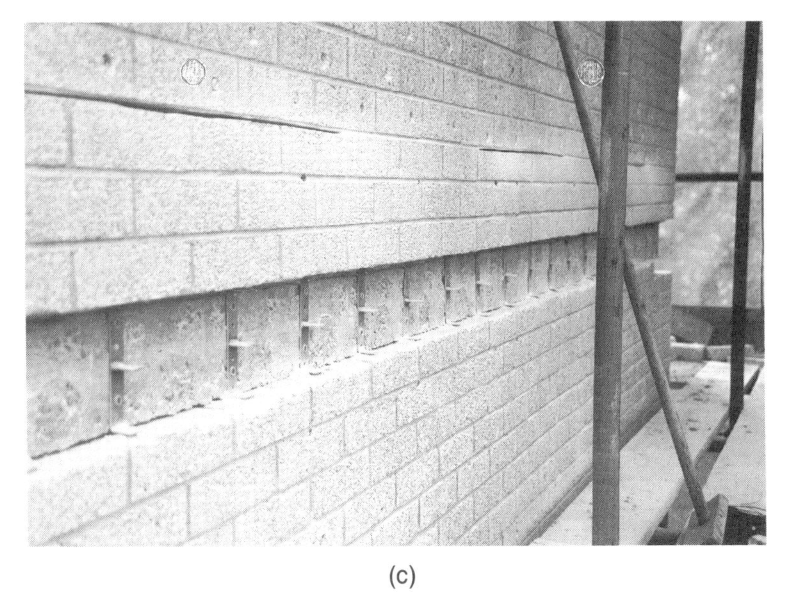

(c)

(a)

(b)

Fig. 14. (a) Panel wall failure; complete wall moved under wind suction; return end shows gap between wall and window frame; (b) movement of wall at top of building; (c) whole panel of wall shifted on DPC; (d) excess cavity width reveals shortage of ties, short ties and ties pulled out from blockwork; walls were largely resupported at intermediate levels, parts were rebuilt and remedial ties were inserted over retained portions

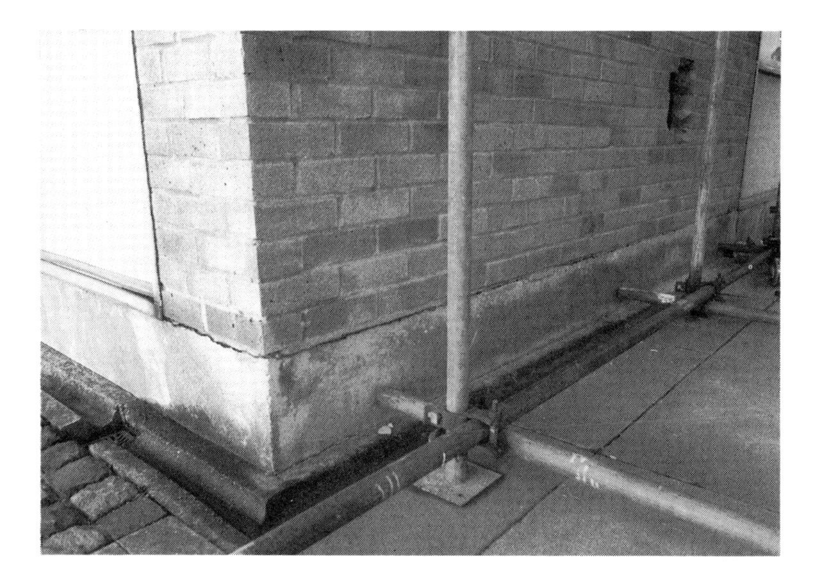

(c)

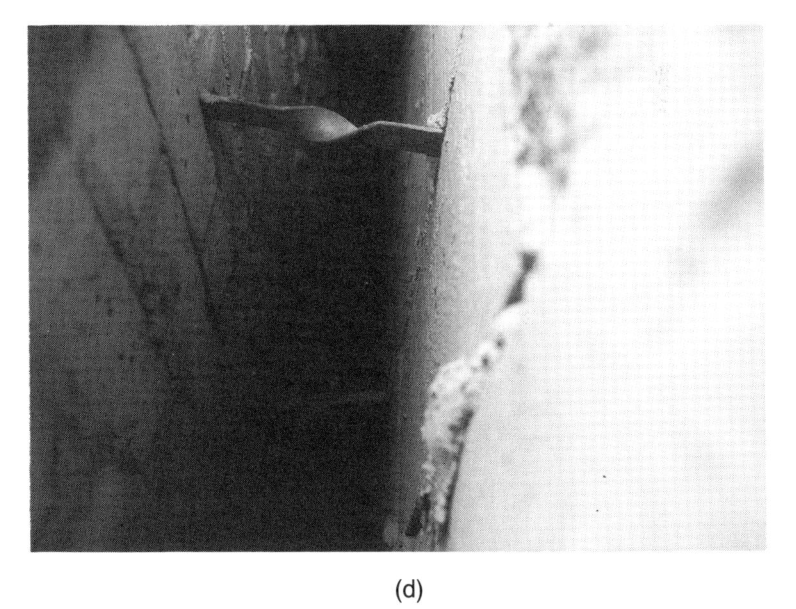

(d)

9

Design of claddings and fixings

9.1. Introduction

The majority of modern buildings are clad with an outer layer of weatherproof material which may be considered the climatic overcoat. Invariably, these outer layers are non-load-bearing, yet by their very nature they need to provide a weatherproof environment, and at the same time reflect every movement to which any building is subject.

Depending on the particular material, independent support needs to be provided and restraint to the main structure or backing material achieved. The important point to bear in mind is that once in position the fixing becomes inaccessible, and therefore attention needs to be given to the type of metal used in the manufacture of the fixings.

Cladding can take many forms, and can be functional or decorative or both. The most common forms of cladding are

- precast concrete panels
- facing brickwork
- stone (either natural, terracotta or reconstructed stone)
- various forms of metal sheeting
- slates
- glass curtain walling
- glass-reinforced cement (GRC)
- glass-reinforced plastic (GRP).

The attachment of cladding to the main structural elements of a building has become more sophisticated in recent years, although fixings of a quite familiar nature have been used traditionally in classical buildings for centuries.

The classical buildings of Greece and ancient Rome were constructed with massive stones which were largely self-supporting. Natural bonding of the stones added to the stability, and yet connecting cramps and dowels have been found in the stones.

As architectural styles developed, smaller stones tended to be used and cramps and ties became more common. Over the years stones became thinner and the problems of corrosion became more obvious.

The use of iron cramps coated with oil or bedded in lead was a common practice well known to the Romans. However, as building design continued to evolve and new techniques of construction developed, there became a need for new materials to take the place of iron fixings, which became increasingly subject to corrosive attack. At this time, copper became widely acceptable and was used for natural stone cramping situations. Subsequently, however, the traditional bonded ashlar and brickwork were replaced by simple skins of material tied back to the main structural frame. The stresses experienced became higher and of greater significance; copper, with its low tensile strength, became unacceptable.

The use of copper alloys such as phosphor bronze became more widely used and more suitable for the larger precast concrete panels being developed. There were however drawbacks to some of these alloys. The manganese bronzes were later found to suffer stress corrosion; also, all the materials were, and still are, extremely costly to manufacture. Many items were cast in foundries, and this method of manufacture is extremely expensive and time-consuming. There are now very few traditional foundries in existence, and few cast fixings are now manufactured from these materials.

Stainless steel has rapidly replaced the copper alloys because of its versatility, its relative ease of manufacture and its more acceptable cost. Mild steel is still used with protective coatings of zinc, but its limitations have become all too familiar in recent years, in relation to wall tie failure.

9.2. Selection of fixings

It is clear that the various requirements noted above dictate the basic parameters in the selection of the type of fixing and the material to be used.

- The fixing material must be corrosion-resistant, for once in position it is usually inaccessible and thus impossible to replace.
- The metal should be non-staining if used directly against the retained or supported cladding element, and most especially if the material is vulnerable to discolouration.
- The metal should have sufficient strength to resist the applied loads with economy of section.
- The metal used should normally be specified from standard material thickness and width. Non-standard materials, apart from difficulty in purchasing, are also costly.
- The metal should have good workability and should retain its strength under working during manufacture.
- Cost is important in the context of design requirement, but it should be noted that price is not always of prime importance; integrity and safety should at all times be the main considerations.

Although there are many types of cladding material, there are only two types of fixing: load-bearing (Fig. 15) and restraint (Fig. 16). Load-bearing fixings, as the name implies, support all applied

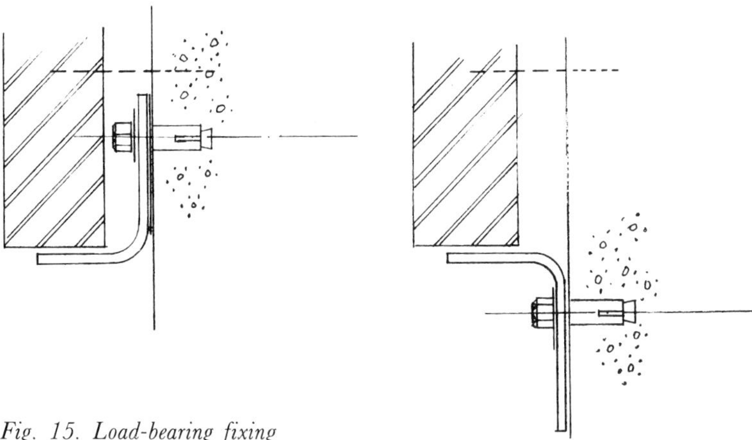

Fig. 15. Load-bearing fixing

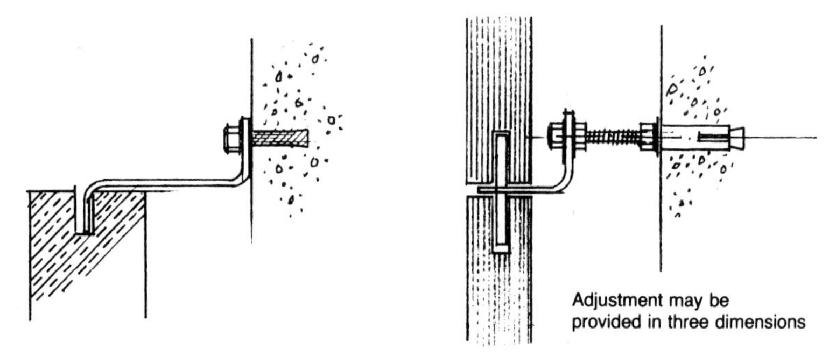
Adjustment may be
provided in three dimensions

Fig. 16. Restraint fixings

structural loads, including any superimposed loads, services and secondary fixings such as doors or windows. Restraint fixings take all temporary applied loads such as wind action (in all directions) and any out-of-balance restraining forces due to eccentricity of the cladding member (large overhanging cornice stones or similar). In cases of very high out-of-balance forces, it is often prudent to consider this type of fixing as a load-bearing element. Load-bearing fixings can, of course, be used to restrain as well as to support.

In the design of fixings, there are two distinct stages: the method of attachment to the cladding material; and the method of attachment to the main supporting structure. It is important that the location of the fixings should allow the loads to be transferred from the cladding material to the supporting structure without overstressing the cladding locally, the fixing itself or the backing material.

9.3. Development of fixings

Over the past 35 years fixings have developed in importance and sophistication. Many of the traditional methods of fixing stone cladding, especially marbles and granites, are no longer acceptable. Where once it was acceptable to 'fix' stonework panels with mortar dabs and perhaps wire ties, particularly externally, the trend towards the modern high-rise buildings dictated the need for more positive retention. Facing stones are expensive, especially polished marbles and granites, and when mortar is placed behind stones there is a tendency for the salts to leach through and cause discolouration on the face of the stone without special treatment at the rear face of the cladding. No client is willing to accept this form of environmental degradation to prestige buildings.

Incidents of failure have become more common but what is more worrying is the tendency to seek restitution through the courts. This dictates that far closer attention needs to be given to the analysis of the fixing design and more stringent attention to site supervision.

The arithmetical calculations are becoming more onerous and the fixing designer is required to have a much wider knowledge of all aspects of building design and materials; his responsibility has increased commensurately. Building movements, jointing materials, forms of protection, and methods of manufacture all need consideration. By the same token the fixer's job has changed, and the modern fixer needs a wider knowledge of fixing techniques than his predecessor. Unfortunately, the traditional masons' skills are to a large extent being diluted. Modern methods of cutting, planing, turning, morticing and polishing stone have taken much of the need for traditional skills away from the old masons.

9.4. Philosophy of design

The type and size of a fixing will vary depending on the type of building cladding used. However, the basic approach for all materials, with the exception of normal bonded brickwork, will be similar. The size of the fixing will necessarily vary according to the weight of the material.

Precast concrete is the heaviest form of cladding unit and if movement occurs it has a serious effect on the fixing attachment. Stresses developed by thermal movement can have devastating effects on fixings, especially at the point of attachment to the main structure. Several failures investigated have consisted of the breaking out from the concrete backing structure of the cast-in-channel attachments. Restraint straps and angles may also be badly deformed by the movement of heavy panels.

The fixing of GRC and GRP panels is usually similar to that of precast panels. However, there is one important difference: precast panels, if adequately reinforced, can be suspended from top fixings; GRC and GRP must be bottom-supported.

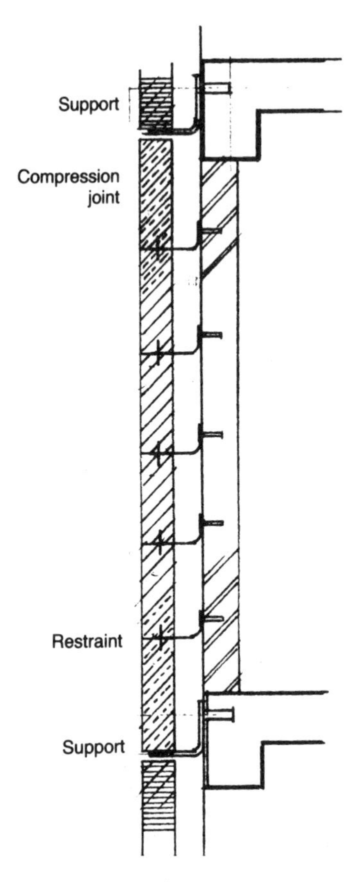

Fig. 17. Standard support and restraint

Natural stone includes marble, granite, slate, limestone and sandstone. The term thin facing applies to marble, granite and slate, which will be cut 20 and 30 mm thick for internal and low-level use and 40 mm thick for all conditions above first floor level. Limestone identified as hard limestone, i.e. travertine, will also fall into this category. Limestone such as Portland and sandstone will be cut 50 mm thick for internal and low-level use. For all conditions above first floor externally, it should be 75 and 100 mm thick.

Panels of thin facing stones would normally be supported at each floor level or similar suitable points with the individual stone retained; however, it may be economical to support separately each individual stone. This latter method is more widely adopted in other countries, especially in Europe and the Middle East. It offers considerable aesthetic advantages, but requires a concrete structure rather than the traditional column and beam construction common in the UK.

The heavier stone facings are again supported at suitable levels but detailed design often dictates separate support points, as is shown below in the more detailed design analysis (Figs 17 and 18).

Normally brickwork, an integral bonded material, will be supported at the floor levels and retained with wall ties to the inner skin (if a cavity wall), or the backing wall. The spacing of the wall ties is dictated by the requirements of the relevant British Standard. Brickwork is generally supported on continuous stainless steel angles. All facing materials will need to be jointed horizontally and vertically and such joints must coincide with any structural expansion joints. There will, of course, normally be more joints

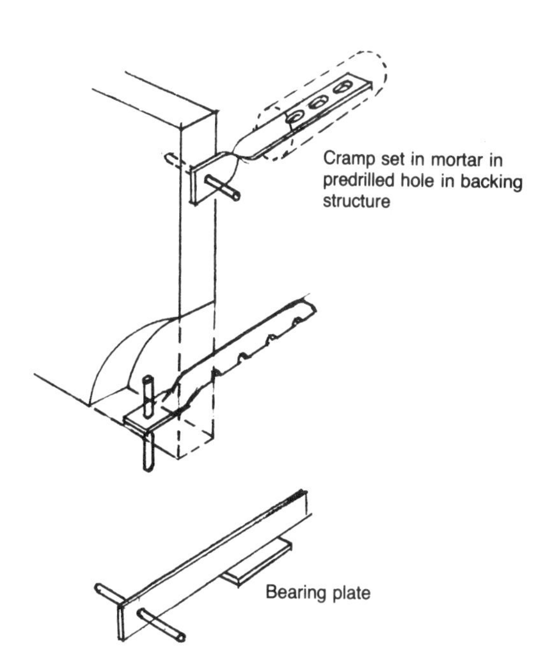

Fig. 18. Typical load carrying/restraint cramps: generally European

in the facing materials than will be necessary in the main structural elements.

One cardinal rule to observe is that any cladding element such as stone, precast concrete or GRC should be supported at only two points. The introduction of a third support point would invariably lead to unevenness, and would cause unbalanced loading and stress in the panels. Loading should be equally divided between the support points whenever possible. Alternatively, restraints should be used to take out eccentricities (Fig. 19), as indicated in the detailed designs below. Individual stones should be restrained at four points (Fig. 20).

In the design, consideration should be made of additional building movements such as elastic compression of the structure and structural deflections of edge beams. The locations of cladding units relative to the compass direction and environmental effects are important.

The most vulnerable material in this respect is GRC. In the early days of its use advantage was taken of its lightness in weight and flexibility in manufacture to allow complex shapes to be cast. This ultimately proved disastrous in the case of many cladding sandwich panels. The panels were often painted in dark colours; thermal movements were large and deformation caused excessive cracking of the panels and movement of the fixings, causing break-out from the panels.

Other environmental conditions applicable to all cladding elements may be broadly classified with respect to atmospheric pollution. These range from rural conditions to severe industrial situations. Consideration must be given to rainfall, humidity, temperature, smoke and chemical pollution. Location within certain buildings can lead to problems with condensation, and

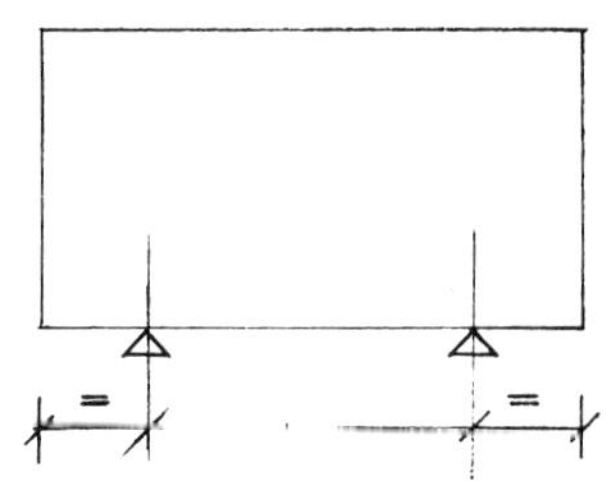

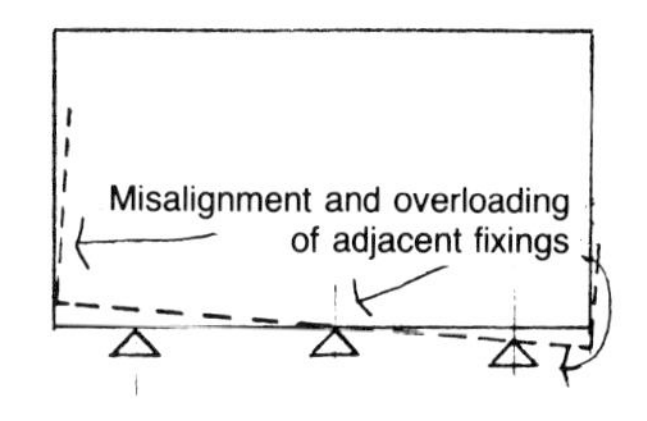

Fig. 19. Support requirements

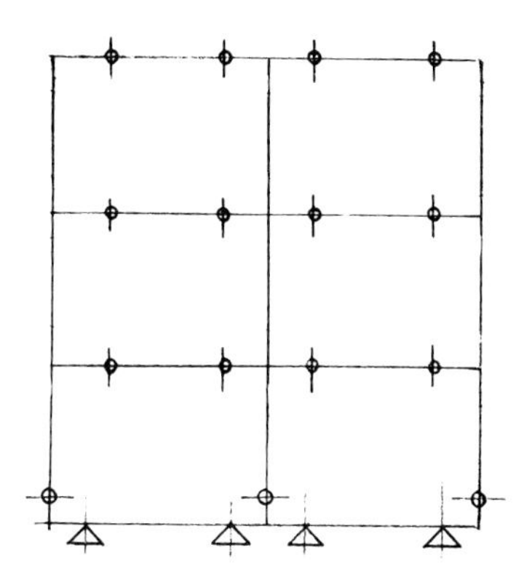

Fig. 20. Restraint location

this is particularly difficult in buildings such as swimming pools. Marine environments are another special area requiring close attention.

9.5. Site realities

A good fixing should be sufficiently flexible to accommodate reasonable site discrepancies (Fig. 21). Failure to provide a means of tolerance in a fixing can lead to an awkward or misaligned fixing, or even in some cases to the complete omission of a fixing. Such situations have proved to be a contributory factor in many cladding failures.

9.6. Materials

Brickwork

In the context of support, the type of brickwork considered has little effect. However, it is important to consider density and absorption and also the type of mortar in which the bricks are bedded. This latter factor may have a very dangerous effect on the metal ties used in conjunction with stability and retention.

Normally, fired clay masonry units should comply with BS 3921 or the recommendations of DD34. Calcium silicate bricks should comply with BS 187 or the recommendations of DD59. Concrete

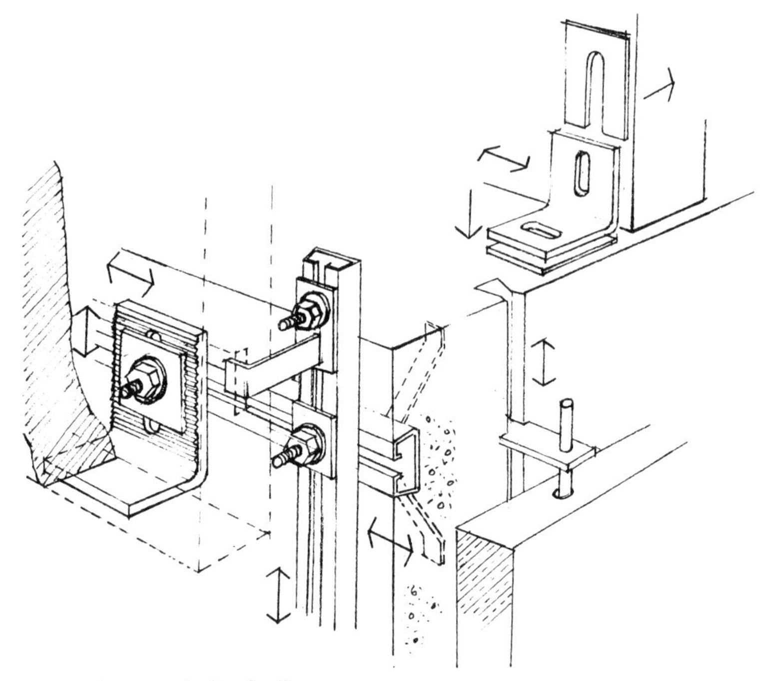

Fig. 21. Some methods of adjustment

masonry units should comply as follows: concrete fire blocks — BS 2389; precast concrete masonry units — BS 6073 : Part 1; reconstructed stone masonry units — BS 6457.

The density of the bricks is important with respect to any expanding bolt fixings which may be contemplated in the fixing design. Weight obviously has an influence on the support load calculated.

Stonework

The choice of a building in stone is usually made on the basis of the aesthetic and environmental requirements. However, technical consideration should be given in relation to the methods of fixing. There are three primary groups of stone: igneous, sedimentary, and metamorphic. Igneous rocks are formed from molten magma of volcanic origin; they include granite, basalt, and diorite. Sedimentary rocks are the result of the disintegration of rocks by weathering and erosion; the resultant deposit is subject to the action of water or air which redeposits them, and the layers eventually harden to form rocks or clays. The principal sedimentary rocks are limestone and sandstone (these rocks have many subdivisions). Metamorphic rocks are rocks which have undergone a change due to pressure or heat; the principal stones in this class are slate and marble.

Cast stone

A common building material grouped with stones is cast or reconstructed stone. This is manufactured from cement and natural aggregates to give the appearance of stone. Common items of manufacture included in this range of products are sills, lintels coping stones and ashlar.

Reinforced concrete precast units

These units do not require description; they should, however, be considered in two respects. Precast concrete units can be structural supports as well as decorative cladding panels. In either case, the fixings required will be large owing to the weight of the units and more than in any other material consideration needs to be given to methods of erection as well as fixing.

GRC (glass-reinforced cement)

This is a material favoured for its flexibility and low weight. Units are made to have the same external appearance as concrete cladding units, but because they are lighter the tendency has been to over-size and complicate the shape of the units. The fixing requirements are similar to those of precast concrete, but of lighter weight and smaller dimension.

GRC is manufactured in two ways: the premix process and the spray up method. In the premix method the glass fibres are simply mixed to give a random arrangement of fibres. In the spray up method the fibres are chopped and sprayed simultaneously with the fluid matrix to give a more even two-dimensional arrangement of the fibres.

9.7. Design criteria

The selection of fixings must be considered an integral part of the cladding design, and the method of attachment of one to the other must be compatible. The basic principles to be considered in fixing design are as follows

- the form the main structural elements take and the material of the structure
- the type of cladding material used
- the design forces to which the cladding and the structures are subject, and any incompatibility or out-of-balance forces; this will largely dictate the type of metal to be used and the size and number of fixings
- the characteristics of the metal, considering its workability, availability and compatibility
- the required life of the structure and the possible aesthetic requirements with respect to the visual appearance of any exposed fixings; it should be noted here that expediency and cost-saving should never take precedence over integrity.

There are many methods available for attaching the cladding material to the fixing (Fig. 22). When considering precast units or manufactured items of any kind, the fixing may be cast in. These items include cast-in sockets, cast-in channels or dovetail slot, and cast-in metal attachments. In lighter or natural materials

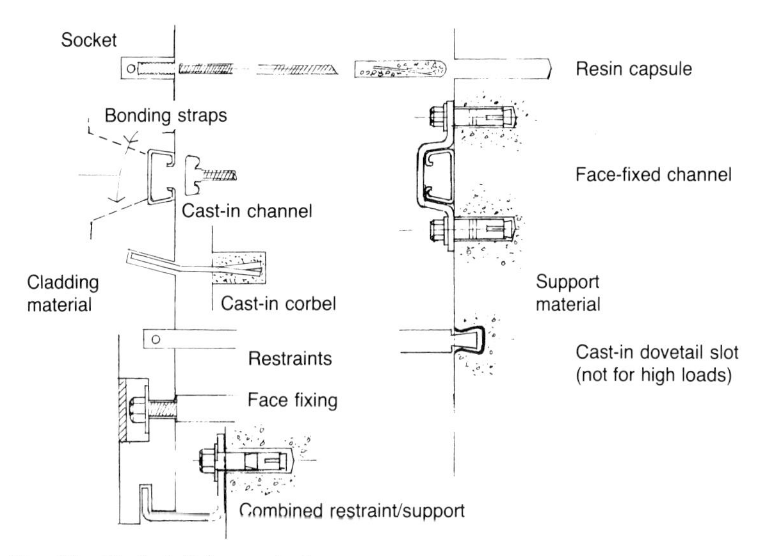

Fig. 22. Typical fixing methods

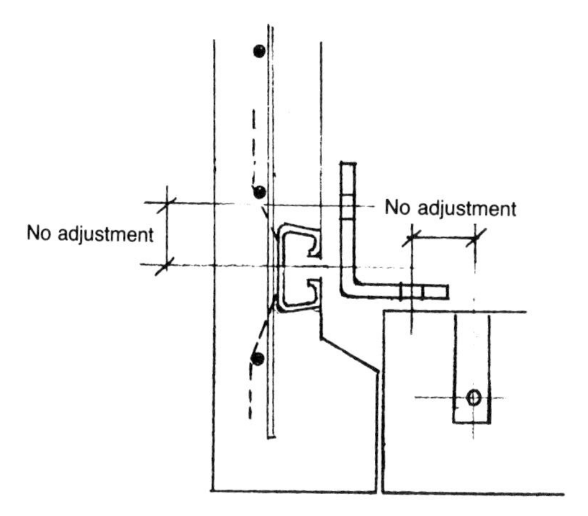

Fig. 23. Precast unit tolerance

there are dowels, ties (set in mortices), plates set into mortices either bolted or grouted in, through or face fixings, and lipped or nibbed angles set into mortices.

The methods of attaching the fixing to the main structure are also various. The type often depends on the stage at which the cladding design is considered. Unfortunately, the cladding considerations are very often left until after the main structure has been constructed. This predetermines that the fixings need to be of a post-attached type; these fixings include

- drilled in expansion bolts
- resin anchors
- injected grouted fixings
- mortared in ties, cramps or corbel plates
- surface-mounted channel fixings.

If it is possible to use pre-positioned cast-in fixings, these include cast-in channels and cast-in sockets.

It is essential to note that the positioning is of the utmost importance if cast-in fixings are to be used. The casting-in of fixings where in situ concrete work is concerned is extremely risky, and only if strict control can be exercised should this method be adopted.

In the manufacture of precast units the control exercised may be much more easily supervised, and this is a reasonable approach to adopt. It must, however, be considered in the light of site realities, as previously mentioned. It would be foolhardy to consider cast-in fixings in both structure and cladding unit if due allowance were not made for tolerance (Fig. 23).

Similar consideration needs to be given to post-drilled fixings, especially if the supporting structure is a heavily reinforced concrete frame. The drill positions may coincide with the reinforcement, and only in exceptional situations should an engineer allow his reinforcement to be cut; therefore toleranced (slotted) holes should be used where practical, always remembering that a vertical slot should never be applied to a load-bearing fixing unless some form of anti-slip connection is applied, such as serrations on plates and washers. Nevertheless, the principle of post-drilled fixings gives considerable freedom within any design constraints which may apply. The efficiencies of the various types of fixing will vary, depending on whether the concrete is dense or lightweight.

9.8. Positioning of fixings

There are two important points to consider in positioning fixings: they should be positioned so that the retained panel cannot rotate, and they should be positioned so that all fixings are equally loaded or at least designed to take out eccentricities (Fig. 24). In both

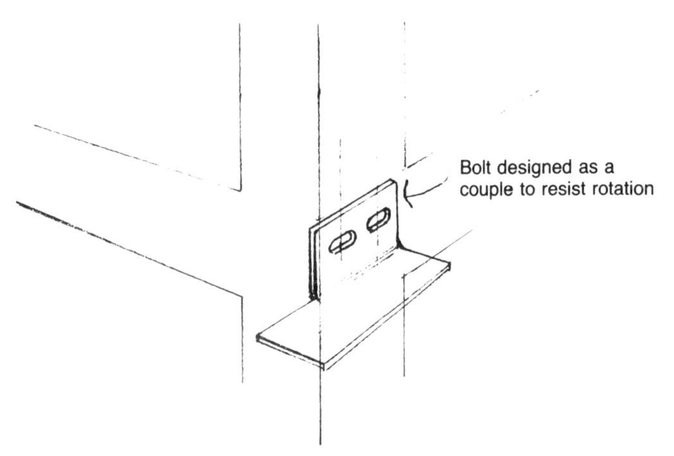

Fig. 24. Corner support: typical eccentric corner bracket

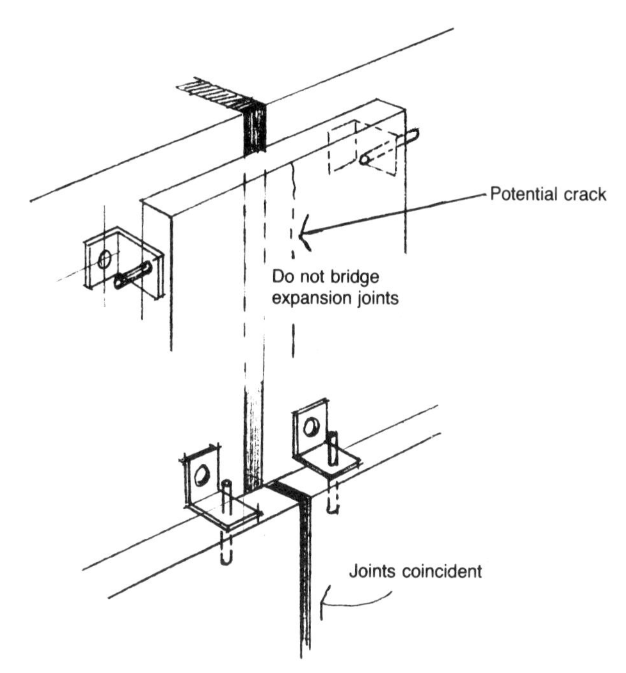

Fig. 25. Expansion joint treatment

cases it is clear that the method of attachment should be such that the cladding element, the structure and the fixing itself are not subject to over-stress in any combination of load.

Supported units should normally only take bearing on two points, and these points, wherever possible, should be equally loaded. The introduction of more points may easily cause an uneven distribution of load, as it is not easy to guarantee three points at a similar bearing level.

Restraints should normally be placed at four points, again disposed to give equal loading. In large precast units of concrete or GRC, provided the element is restrained in the recommended manner, the units need not be located together.

In masonry facings such as natural stone or precast masonry the elements may be located together, and each may be restrained to the structure. In locating elements together, due notice must be taken of compression, movement or expansion joints; such joints should not be bridged unless this is absolutely essential

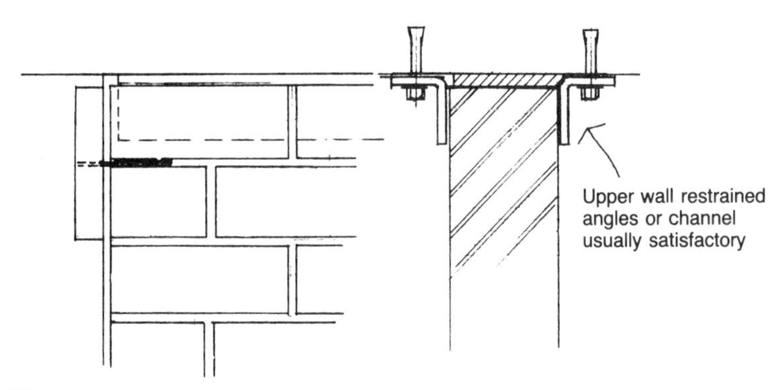

Fig. 26. Brickwork restraint

(which is not frequent). If they are, the fixing should be capable of movement within the joint.

Thin facings such as marble and granite may be treated as for heavy stone masonry, except that the fixings are generally lighter (Fig. 25).

Brickwork is generally of two types: complete facing, and a panel within a structural frame. In the latter case the panels should incorporate movement joints, most importantly at the head of the wall beneath the beam soffit, and sometimes down the sides. However, the wall panel must be restrained and the restraint fixings must be able to provide movement and the required degree of restraint (Fig. 26). Any wall panel sitting on a damp-proof membrane cannot be considered restrained and certainly should not be calculated as a cantilevered wall.

Restraints should not pass through any damp-proof course; however, if they do the damp-proof material should be suitably sealed or made good. In the case of any wall panel, brick ties must be used at the required centres as laid down in the relevant Codes of Practice. When considering a complete facing skin, vertical movement joints and horizontal compression joints should be incorporated as necessary; wall ties should again be used at the stipulated centres.

The method of attachment of the tie to the structure depends on the structural backing material, and there are numerous permutations.

When designing a restraint system using four points of retention located into adjacent stones, care must be taken to ensure that the following adjacent or upper course stones may be located (Fig. 27). If fixings are used in bed joints, side fixings cannot be used in the same stone, as the positioning of an adjacent unit cannot be made onto locations in a vertical and horizontal direction simultaneously. The only time at which such an arrangement may be made, but only with the side (horizontal) location pointing into one stone, is at a top course of stones, immediately under a compression joint. The 'loose' fourth corner of the unrestrained stone may usually be located to its adjacent stone by using a dog cramp or wire between two stones on the top.

With bonded ashlar walling the situation is much more flexible, as the fixings in any course may be reversed in direction as required. The most difficult fixing is the last one in a row at the end of a run of stones, especially if the stones or panels are recessed into a frame. This fixing, if badly designed, is often left out on site, and this may lead to a local failure.

A frequent cause of trouble on site is dimensional discrepancy. The fixing must be sufficiently mechanically flexible at both the cladding location and the structural fixing face to make location easy, but the whole fixing should be designed to take up construction tolerances and/or discrepancies in unit and structure. Careful thought in design may eliminate the danger of an awkward fixing being omitted, and this will ensure continued site progress.

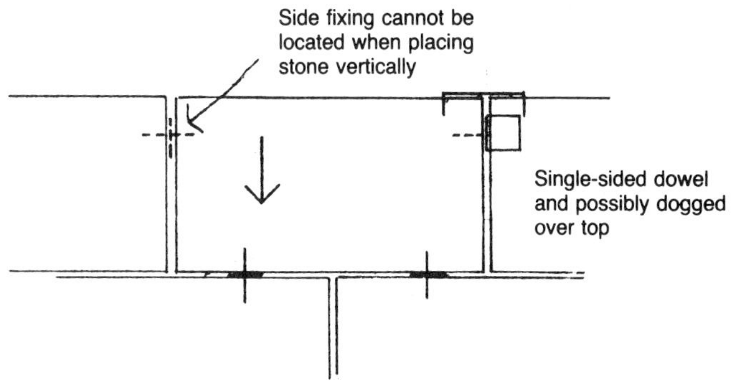
Side fixing cannot be located when placing stone vertically
Single-sided dowel and possibly dogged over top

Fig. 27. Restraint difficulties

In general, it is good practice to position the fixings as the cladding proceeds; this helps to ensure accurate positioning of the fixings.

A fixing into a structure should always be positioned above the line of the unit to which it is to be attached, in order to make fixing, measurement and location as precise as possible (Fig.28). A fixing at the back of a stone, when erected, cannot be easily adjusted, nor can the stone easily be located unless loose dowels are used.

It is good policy and good design to keep the number of components in any fixing to a minimum. Loose dowels, nuts and bolts, though not always avoidable, may become lost or dropped from scaffolding. This is a potentially hazardous situation in that the fixing may be omitted or left only partially located.

9.9. Design loading

Loadings are of two types, as in any structural design

- dead load: this is essentially the total weight of the unit or units being supported, and it is easily calculated as the material density multiplied by its volume
- live load (or applied load): this is not always applicable to cladding situations as an applied live load, but it does include

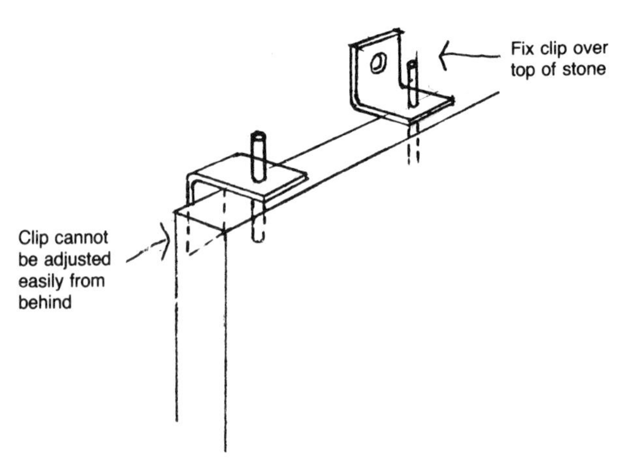

Fig. 28. Locating restraints

applied material loads, services, windows and other secondary fixing loads.

Also to be considered are expansion forces, vibration, impact, etc.

Wind loads
These loads can have a very important bearing on fixing design, and in severe cases they should be considered from the cyclic effect likely to be applied to any restraint fixing in softer backing materials such as blockwork. Wind loads may act in virtually any direction and may have a positive or negative effect on the fixing.

Special considerations
Vibration, impact and explosion should be taken into account in the design, but owing to their respective complexities each should be considered on its merit.

Earthquake forces
The restraint fixings should be designed to accept horizontal forces caused by earthquakes, and should resist forces which are at least equal to the dead weight of the unit being fixed.

Movement of structure
Apart from the localized thermal or moisture movement, the overall total effect on the structure of earthquake, subsidence (natural and man-induced, i.e. mining), and any other form of geophysical movement must be considered. Such general movement considerations will normally be allowed for in the main structure design. However any expansion, differential or settlement joints must be reflected in the cladding units, and the cladding must never pass across a joint without some allowance for similar movement (Fig. 29).

Even the smallest of dimensional movements can have a devastating effect on cladding. It is important to mention that any type of movement joint should be properly formed, and the appropriate joint filler and sealing compounds should be used strictly according to the manufacturer's recommendations. On no account should loose mortar or spacer blocks be allowed to remain in the joint, and the designed joint thickness must be maintained in order to allow the free and correct movement required.

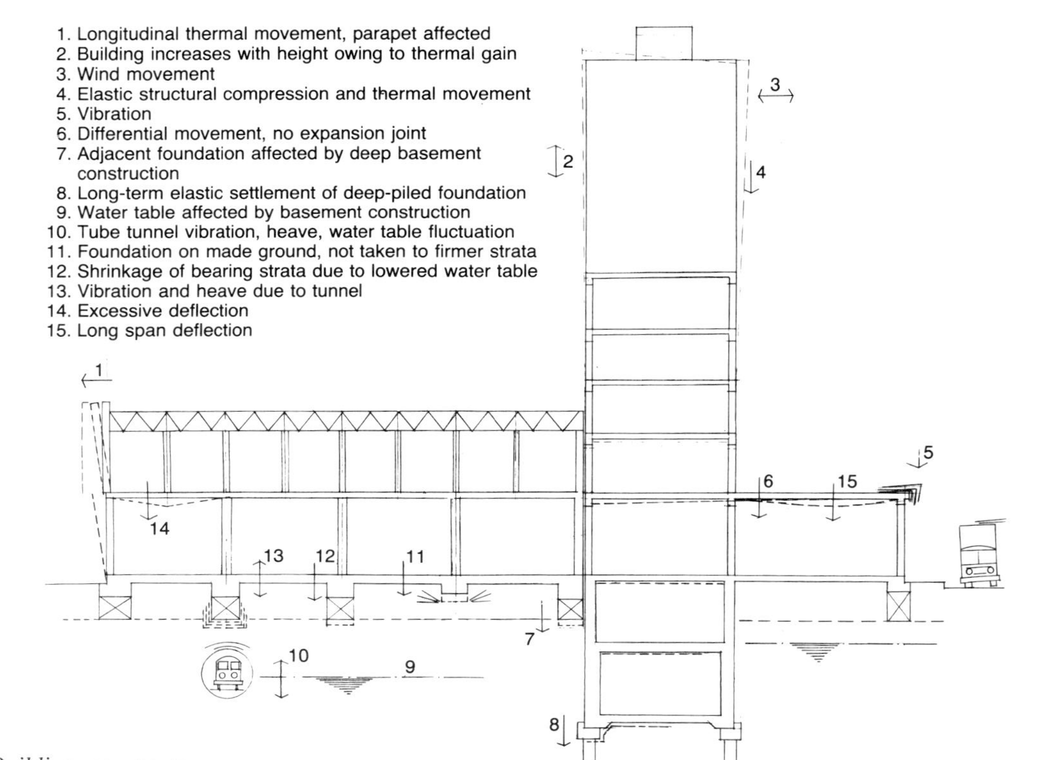

1. Longitudinal thermal movement, parapet affected
2. Building increases with height owing to thermal gain
3. Wind movement
4. Elastic structural compression and thermal movement
5. Vibration
6. Differential movement, no expansion joint
7. Adjacent foundation affected by deep basement construction
8. Long-term elastic settlement of deep-piled foundation
9. Water table affected by basement construction
10. Tube tunnel vibration, heave, water table fluctuation
11. Foundation on made ground, not taken to firmer strata
12. Shrinkage of bearing strata due to lowered water table
13. Vibration and heave due to tunnel
14. Excessive deflection
15. Long span deflection

Fig. 29. Building movements

9.10. Selection of metals

The selection of a suitable metal for any type of fixing depends on several factors; basically, they are availability, suitability and workability. The method of manufacture (fabrication) may also have an effect on its strength.

The materials most commonly used at present and their principle areas of use are set out here

- loadbearing fixings
  - phosphor bronze
  - aluminium bronze
  - silicon aluminium bronze
  - stainless steel
- restraint fixings
  - phosphor bronze
  - aluminium bronze
  - silicon aluminium bronze
  - stainless steel
  - copper.

In certain controlled applications varieties of these metals may be used.

Iron

Iron is an element found in rocks in the form of ore; only those ores containing 20% or more of iron are considered to be iron ores. The three principal ores are magnetite, haemetite and limonite.

After preparation of the ore by crushing it is converted into crude iron by smelting to form pig iron. During smelting iron picks up several impurities such as carbon, silicon, phosphorous manganese and sulphur. High phosphorous ore gives iron known as 'basic', but low phosphorous ores (haemetite) give iron known as 'acid' iron.

All the elements found in pig iron affect its properties, but carbon is the most important, appearing in quantities up to 4%. The carbon is either in combined form giving cementite, or as a free element (graphite) in flake form. The form that the carbon takes in iron is especially critical to the manufacture of stainless steel.

Steel

Steel is a term widely used, but not fully understood. As stainless steels of specific grades and type are generally recommended, a brief explanation of this will be of use.

Steel is an alloy of iron and carbon, and is classified as carbon or alloy steel. Grades of steel are

- low carbon (0·0–0·1%)
- mild steel (0·1–0·33%)
- medium (0·34–0·6%)
- high carbon (0·6–0·9%)
- tool steel (0·9–1·3%).

The ingots of steel obtained from the various manufacturing methods are worked by rolling or forging. Hot rolling is preferred, as it is beneficial to the crystalline structure, and hot forging avoids high concentrations of stress. Cold working causes crystalline deformation, and cold rolling imparts considerable hardness which has to be rectified by annealing if required.

The crystalline structure of the steel is important and the proportions in which the carbon is contained in the steel is governed by temperature. There is a complex relationship between iron and carbon.

Carbon can exist with iron, making cementite, or as a separate substance known as Phase. Iron is also an allotropic element, which means that it can exist in more than one form: carbon, diamond or graphite. The control of the carbon/cementite and pearlite structures of steel is achieved by a combination of temperature variation and cooling techniques. These qualities help to form the various types of stainless steels, the three main types of which are martensitic, ferritic and austenitic. The addition of alloying elements with steel forms the stainless steels, of which the principal element is chromium.

The stainless steel used for structural purposes is the austenitic variety, and the other alloying elements used are

- manganese
- molybdenum
- vanadium
- tungsten.

Manganese at 15% stabilizes the structure at normal temperature so no change takes place during cooling. The resulting steel is non-magnetic and has pronounced work-hardening properties able to withstand intense wear. Molybdenum added at 0·5% avoids temper brittleness. Vanadium gives high tensile strength, and tungsten gives hardness at red heat.

Aluminium

For many years this metal was difficult to produce, because it was impossible to extract in quantity owing to its affinity for other elements, particularly oxygen. However, modern electrolytic reduction processes enable the metal to be extracted more readily from the bauxite ores in which it is found.

The principle method of manufacture is to crush the bauxite and digest it in a hot solution of caustic soda. This dissolves the aluminium as sodium aluminate and impurities are separated by filtration. In the electrolysis of the oxide a large direct current is passed through the melt at around 1000°C. The principal properties of the metal are its low weight (approximately one-third the weight of steel), its malleability and its ductility. It is a good conductor of heat and electricity, and may be cast, machined, forged, soldered, brazed, welded, drawn, extruded, etc.

When alloyed with other elements its strength may be increased up to and above that of mild steel. Common aluminium alloys are

- daralcium containing copper, manganese and magnesium
- aluminium/copper/nickel (or Y-alloy)
- aluminium/magnesium, with magnesium, manganese and silicon
- cast aluminium alloys containing zinc and copper
- silicon giving a fluid mix suitable for intricate castings.

Copper

This is one of the most important elements; it is mined in many parts of the world. The most common ore is pyrites, which contains around 34·5% copper. The ore is crushed and the dust is separated by flotation. The concentrate is smelted with a flux. The resultant matte of copper and iron sulphide is heated in a Bessemer converter, followed by refinement in a furnace or by electrolysis using sulphuric acid and copper sulphate.

The copper alloys include brass, bronze, aluminium bronze and cupro-nickel alloys. Brass is a term which covers a wide range of copper zinc alloys with varying properties and good corrosion resistance. Bronze is a term which covers a wide range of copper tin alloys, including the phosphor bronze used in engineering castings.

Aluminium bronze is used in marine and general engineering applications where high strength and hardness are required, along with high corrosion resistance. Manganese bronze contains manganese, tin, iron and aluminium.

Lead

This metal is used largely as an alloying element in other materials to increase machining qualities. A high lead content makes welding difficult.

Zinc

Zinc has excellent corrosion-resisting properties, and the most important single use of zinc is in galvanizing or zinc coating. The advantage of zinc as a coating element is that it becomes the sacrificial element in the presence of the corrosive solution. Coatings may be applied by either the galvanizing or sheradizing processes.

Galvanizing consists of immersing the article in a bath of molten zinc at above 800°C. The sheradizing process consists of packing the item in zinc powder and heating to a temperature below the melting point of zinc. The attraction between the zinc and iron causes diffusion of the zinc into the iron.

9.11. Fabrication

Fixings for all forms of cladding are generally converted from strip, coil or sheet, or wire, rod and bar. Bolts and other similar elements for anchorage are made largely from rod. These are generally special items made in volume by specialist manufacturers.

Normal support brackets, cramps, ties and similar standard items are sheared, pressed, stamped or fabricated, apart from special requirements, they are made in volume from standard material sizes and thicknesses. It is important to understand that wherever possible standard items should be used, and leading manufacturers produce a range of fixings of various dimensions suitable for most normal fixing requirements. The standard channel sections used for cast-in fixings or strut systems and supports are manufactured by a rolling process.

Many of the fabricated items require welding, and this is a field which should only be tackled by qualified and certificated operatives. Stainless steel is particularly difficult to weld, and only the correct materials should be used. Site welding is not recommended so far as stainless steel is concerned.

Castings are used for some smaller fixing components, but sand castings used for some of the more exotic metals are liable to develop flaws.

An important point to consider in load-bearing fixings is that of size limitation. Most support brackets and long angles are formed, that is, pressed to shape from strips sheared from plates. There is a manufacturing limitation of around 3 m, due to the length of the press. Also, the radiused bend which is common on this type of formed angle has a limitation of about $2t$ (where t is thickness). Smaller radii may be produced, but they are not recommended for use, as cracking or splitting can occur on the bend.

Work hardening also occurs when stainless steel or any other metal is bent, and care must be taken, especially with welding at forming points, to ensure that the steel does not become brittle owing to the application of heat.

Normalizing may sometimes be necessary in practice. Welding fabricated sections can also cause distortion, and although attempts may be made to hold the component parts in the correct position, this can cause undue stress to be locked up in the final workpiece.

It may sometimes be an aesthetic requirement to use a more exotic metal, such as one of the bronzes, in order to produce a desired colour or to resist corrosion. There are not many standard sections suitable for fixing requirements, and this usually dictates that the element should be a casting. Only specialist manufacturers should be approached in these circumstances, and, apart from the difficulty of locating a suitable foundry, the cost and size limitations may be prohibitive. Bronze castings usually need to be thicker than a comparable steel member and are very heavy to handle.

9.12. Corrosion

It is generally considered desirable, and in some cases mandatory, to use stainless steel for the basic brackets, ties, cramps, etc., used in cladding support systems. In some cases 'ordinary' mild steel is used in standard sections and galvanized. However, the term 'corrosion' is generally only associated by the layman with steel products; in fact the term has a far wider implication.

The chemical reactions which take place during corrosion are fully understood. What is not understood are the conditions in which the reactions can occur. There are two basic types of corrosion, namely oxidation and galvanic. There is a third kind known as stress corrosion, and this may be caused by the manner in which an article is manufactured or formed. Certain conditions under the two basic processes can cause locked up stresses to occur within the material. These stresses are not immediately apparent, but there are two conditions under which these stresses may suddenly be released: the first is when an outside agent causes corrosion, and when combined with the effect of 'locked up' tensile stresses in the material, sudden failure may occur; the other condition is when the stresses start to reverse with age or work, and this action, when combined with corrosive attack, may also cause failure. Brass and manganese bronze may be susceptible to stress corrosion, and are therefore not suitable for use as fixings.

The basic theory of corrosion is fairly easy to understand. Atmospheric oxygen converts the surface of a piece of iron into iron oxide. Oxidation of other metals may be quite desirable: bronze and copper take on a patina of bluish green, known as verdigris, which in fact forms in itself a protective coating; aluminium behaves similarly and is used as a protective coating. It is only when oxygen is combined with water that severe rusting of steel takes place.

Atmospheric corrosion can, of course, be quite severe when the causes are from polluting agents such as the sulphur dioxide found in industrial environments. When combined with moisture, sulphuric acid is produced; this attacks the metal, and in steel produces ferrous sulphate ($FeSO_4$). This salt, produced during the actual corrosion process, may have an adverse effect on protective coatings.

Galvanic corrosion is a little more difficult to explain, and the conditions under which galvanic corrosion may occur are not entirely predictable. It is established that oxygen and water combine to start the reaction, and water would therefore appear to be the important agent. Yet if iron is placed in distilled water rusting does not take place so quickly. Pure iron is not susceptible to rusting, yet commercial grades of steel are. It has been suggested that an important factor is the non-homogenous nature of manufactured steel when in contact with some gaseous, liquid or solid substance. From these facts there would appear to be a relationship between corrosion and electrolysis.

Metals are listed in terms of nobility. If two metals are placed in a chemical solution an electric current is produced which passes from one metal to the other. When this circuit is complete, one metal (known as the anode) dissolves away, while the other (the cathode) is not affected. This is most important, for not only do ferrous metals ‘rust’ but now there is the possibility of two dissimilar metals corroding. This type of corrosion is known as bimetallic corrosion, and it indicates that care must be taken in using dissimilar metals in fixing locations.

If unlike metals are used together, in certain cases it may be desirable to use inert isolators or insulators. It should be made clear that corrosion cannot be entirely prevented. Protection is only as good as its application, and generally the more expensive the coating is, the better it will serve.

There is a further type of corrosion: ‘crevice corrosion’. This may occur in welded components, and difficult, complex, fabricated components. Stainless steel does not in itself corrode, although damage to the surface may show rusting; however, the surface is self-healing and rusting is not progressive.

Forming marks may also be identified in stainless steel formed angles, which are left by the steel forming tools, leaving traces of mild steel on the surface of the material. This again is not progressive, but it may cause some concern if not understood. Aesthetically of course, such marks are not always acceptable; stainless steel is therefore not an entirely correct term.

Some of the higher grades of stainless steel are not quite so susceptible to this type of damage, and in certain situations, and also potentially aggressive environmental conditions, a higher grade should be specified. Stainless steel is manufactured in this country to the requirements of BS 1449 : Part 2, BS 1501 : Part 3 and BS 970 : Part 1. The grades used in the manufacture of building fixings and other structural components come under the austenitic range. This range generally has good machining and welding properties. The most common grades are 304, 316, 321 and 313. Grade 304 is the most frequently used and 316 is the better for more corrosive situations. Grade 304 is not fully weld-stabilized, but with the use of the correct stabilized filler wires welding under controlled conditions presents very few problems. Tight angles should be avoided in welding, to limit the possibility of crevice corrosion.

The practice of shot-firing fixings through stainless steel will lead to corrosion of the shot pins and failure of the fixing unless a suitable isolator is used.

9.13. Galvanic protection

There are three main methods of zinc protection, as briefly noted above

- hot dip galvanizing
- zinc plating (electro-galvanizing)
- sheradizing.

Hot dip galvanizing
This method is the most favourable and it provides relatively long-term protection, although it should be understood that it is a sacrificial protection in which the zinc becomes the anode. The method entails placing units in a bath of molten zinc; and the end product is a heavy but largely uneven surface. It is suitable for large bulk items or numerous uncomplicated smaller units, but it has some considerable drawbacks: components with holes and threads will require redrilling, and this immediately re-exposes the metal; secondly, some carbon steels become brittle and liable to fracture when hot-dipped after cold working. Steel in this condition must be normalized. In the cold working of certain components, provided the metal thickness is kept at or below 3 mm and bending is controlled to the correct radius, this problem need not be critical.

Zinc plating (electro-galvanizing)
This process is applied to any small complex component. The coating is reasonably ductile, and the components may be bent

after coating without much damage. The precautions for cold-worked steel do not apply in this process.

Sheradizing

This method entails the diffusion of hot zinc dust onto the surface of the component to give a thin uniform coating. It is not always successful with threaded components, as there is a tendency for a build-up of the zinc dust in the thread, making retapping or drilling necessary.

With all types of coating involving zinc protection it must be remembered that the protective coating is liable to damage and chipping, and great care should be taken in handling and assembling pre-protected components. Other methods of protection now exist, including PTFE coatings; these can have a dual use in that certain grades of PTFE form slip surfaces; these may be baked onto metals and they are very hard. Painting is a traditional form of protection, but it is generally applied on site after erection, apart from an initial coat of primer before the product leaves the factory.

9.14. Fixing design

The design of a cladding fixing system is a very specialist field and should only be undertaken by experienced designers familiar with all aspects of structural building design. The design is one of an overall concept and is not limited to the adoption of a single fixing type used unthinkingly throughout a façade.

Failures of cladding elements may occur for a number of reasons; these include poor selection of material, poor fabrication, inadequate consideration of all design factors, and bad site installation. The latter two are closely inter-related. With these points in mind close attention to detail design is becoming increasingly important.

Before commencing any design consideration should be given to a number of major items, some important and obvious, others not so obvious but nevertherless applicable. The list is not endless, but at least if these major points are considered there is a fair chance that the design will be satisfactory

- weight of the material
- size of the element to be supported
- how it is located onto the structure
- design characteristics of the main structure, e.g. strength of concrete
- method of support of the main building
- spacing of the supporting elements (columns, walls, etc.)
- anticipated deflections of the edge beams
- expected extension of the structure under thermal effects
- elastic compression of the structure
- position of the movement joints
- storey heights
- height of the building
- methods of erection
- accessibility to locate fixings
- any vertical loads applied to the unit
- are the units load-bearing?
- horizontal loads (panic loads, etc.)
- wind speeds (or actual design load; note that the overall wind force on a building is not the same as the pressure that has to be applied in design to an individual element)
- any special environmental considerations (e.g. is it on the coast, is it sheltered inland, is it in an industrial atmosphere, etc?)
- the use of the building (i.e. has it any special internal environmental problems, such as computers, corrosive chemicals, vibrating machinery, etc.)
- any special architectural requirements
- any planning restraints
- any particular specified materials
- contractual requirements
- programme requirements
- delivery requirements
- any other relevant information.

The general method of support for any cladding (except precast concrete) is the use of angle bracket supports; these may be either continuous sections or short brackets (Fig. 30). Generally, when using stainless steel the angle is made by forming cut plates on a brake press. This gives a radiused bend. There are few standard

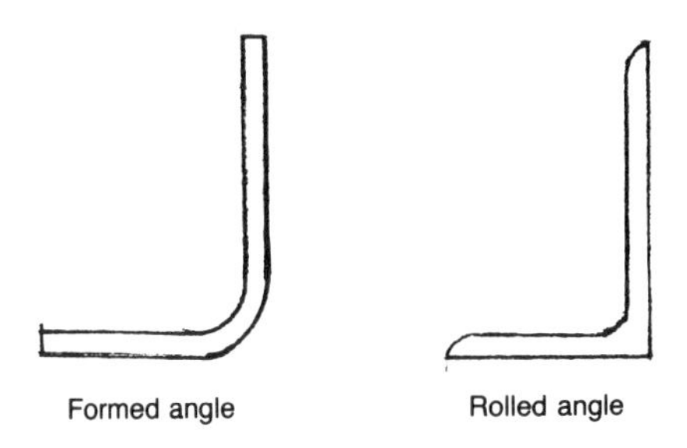

Fig. 30. Types of angle bracket

rolled sections of suitable dimensions manufactured in stainless steel. Mild steel angles of standard British Section may be used if protected mild steel is allowable.

Support brackets may be fixed to the supporting structure by a number of methods, but the most common form of attachment is a bolted connection. The most common forms of these are

- expander bolts
- expanding sockets
- resin capsules
- through-bolt fixings
- cast-in sockets
- nuts and bolts to steelwork
- possible welding to steelwork
- tee bolts in cast-in-channel systems.

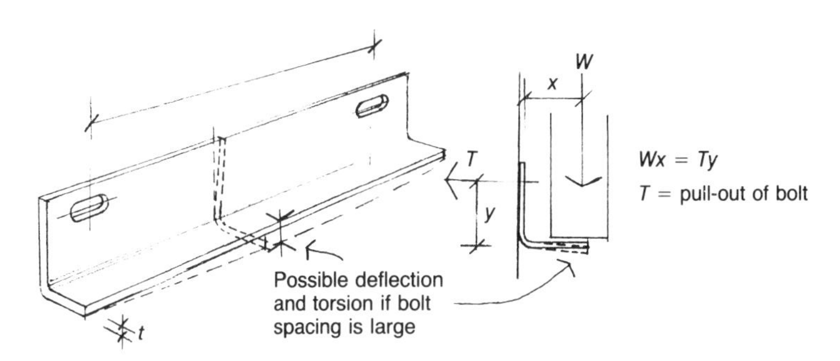

Fig. 31. Long angle design

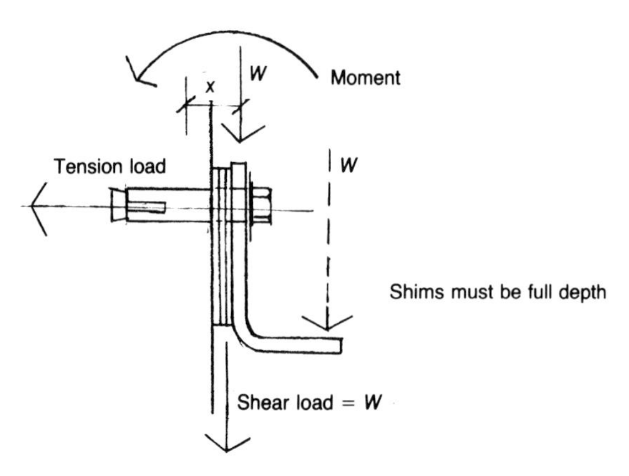

Fig. 32. Support fixing design

The angle as a single element has to support a load, and the design consists basically in checking the thickness of the section as a pure cantilever element. If the angle is long then the spacing of bolts in the angle becomes a consideration, and the angle as a whole 'ell' section needs to be capable of spanning between bolt spacing and of resisting rotation (Fig. 31). The spacing of bolts in this respect is not generally so far apart as to warrant any critical design problem, provided the section size and shape are generally adequate.

The design of the bolt itself is the most important aspect of the design, especially when using expanding bolts. Supporting bolts need to take several types of stress imposed by the action of one load. These are: shear stress, tensile stress and moment stress (Fig. 32).

Tensile and shear stress can normally be combined to give a single principal stress, which may be easily related to an acceptable working stress. However, the addition of a bending stress complicates matters considerably. It is not always possible to predict accurately the distance over which the applied load acts in order to determine a bending moment.

A bending moment is the product of the applied load and the distance between its application point and its anchorage point. In concrete the anchorage point needs to be considered as a point within the concrete, where it is expected that fixity of the bolt will apply (Fig. 33). This is not fully predictable, and it depends on such factors as the length of bolt, and the thickness and strength of the concrete. There is then a further factor which plays a most important part in the overall design. No supporting surface is ever level, and buildings are generally constructed to be within acceptable tolerances, as set out in specifications; these tolerances are generally plus and minus (unlike precision engineering tolerances, which are often plus and zero). This situation means that packing shims of a nominal thickness must always be allowed for in construction (and design) to account for these tolerances.

When a bolt has been designed to resist satisfactorily all these stresses the next most important point is to ensure that the forces can be developed. This is controlled by calculating the maximum torque which needs to be applied to the bolt in order to resist the stresses; then the torque should be applied during installation using a torque spanner.

The torque produces a tensile stress in a shaft which is related to the length, diameter and resilience of the material. Stainless steel loses some of its stress over about a 24 h period after being torqued, and all bolt fixings should be checked after installation. Provided the applied load does not exceed the torque load induced on the bolt, the bolt will be safe. If it does exceed the torque load, there will be a slip.

The other critical factor in the installation of bolts is that of spacing. In the same way as the pressure bulbs on closely spaced piles cause overlap and a reduction in the combined working load, so the expansion anchors set into concrete cause a reduction in the load-carrying capacities of groups of bolts. Most of the recognized manufacturers give guidance in their technical

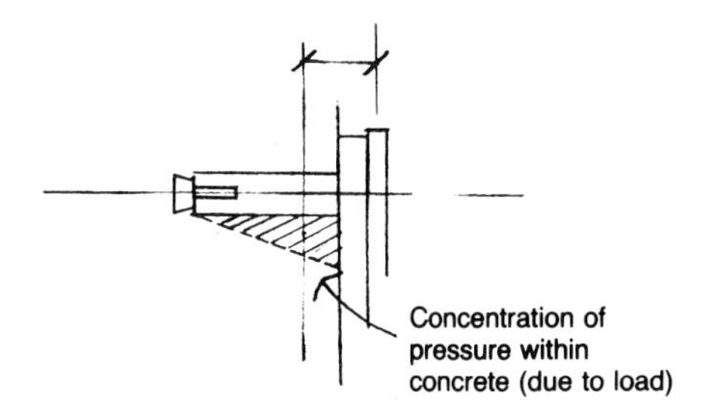

Fig. 33. Stressed area due to bending moment

literature, and there is some discrepancy between various makes. However, it should always be remembered that the published figures are based mainly on laboratory testing, and the engineer is always advised to consider the prevailing site conditions and material strengths before using the maximum suggested design figures.

The other important factor related to spacing is edge distance. Again, the bolt (or any embedded fixing) must be sufficiently far from the edge not to cause a severe and dangerous reduction in the load-carrying capacity. There are various calculation methods, mostly based on test results, which give these additional reduction factors (Fig. 34).

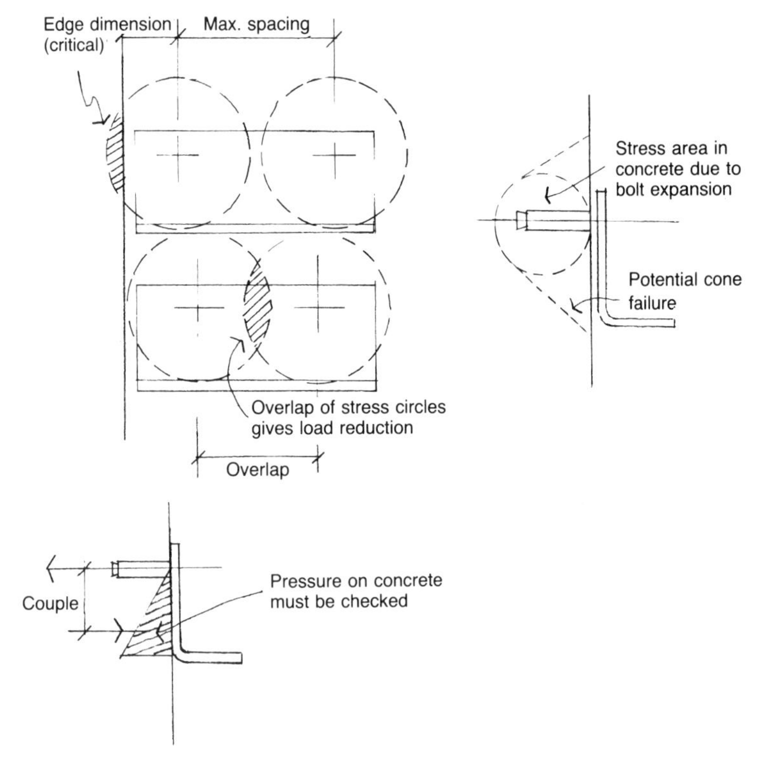

Fig. 34. Bolt spacing recommendations

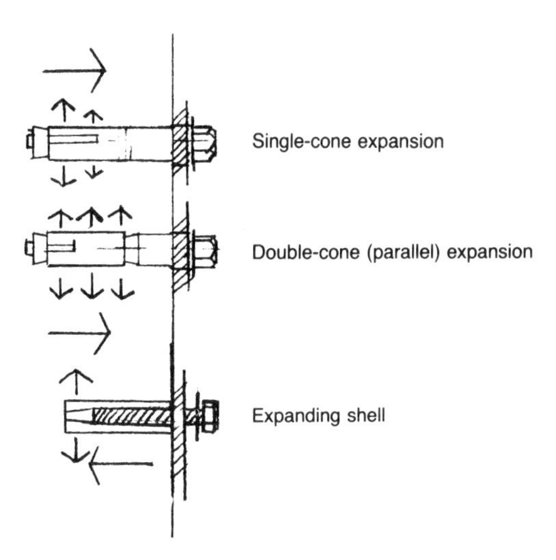

Fig. 35. Bolt actions

One further consideration in expanding bolt anchors is the difference between the two basic types of expansion bolt. The traditional high-integrity bolt, in which a cone is drawn up into the body of the bolt shell, requires the bolt to be torqued to its maximum value to ensure that the shell has been fully expanded to its greatest extent. Normally the applied load will not exceed this figure; if it does, the slip will occur as previously indicated. This second slip is not necessarily dangerous, as the bolt will set further, until the failure load of the anchor is reached. However, once a second set has taken place the clamped bracket is no longer fully tight against the backing material, and overload should be avoided.

The second type of bolt is the expanding socket type, in which the bottom of the shell expands as the bolt is tightened into it. These are not so critical, since they are normally light restraint fixings, and the working torque may be equal to the installation torque, which may not be the maximum value the bolt can take (Fig. 35).

Cast-in sockets are as strong as the material into which they are cast and the manner in which they are anchored into the

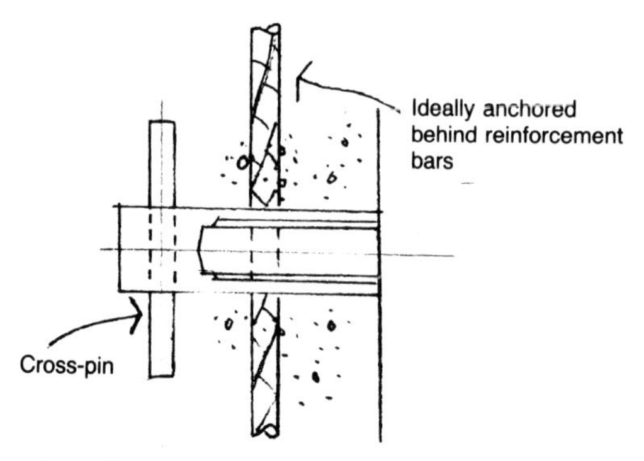

Fig. 36. Cast-in socket

material (Fig. 36). Invariably the actual tensile value of the bolt will govern the load capability, not the pull-out value of the socket. Resin anchors are as strong as the resin and the bond between resin, bolt and parent material into which they are set.

It is important in the installation of any drilled-in fixing to ensure that the hole is of the correct diameter, is drilled straight, and that all dust and debris are blown from the hole before fixing the anchor. Brickwork should not be relied on for high-integrity drilled-in expansion bolt fixings when designing load-bearing supports. Restraint anchors of lower load capacity may be used, with care. Resin anchors or grouted fixings may be suspect in brickwork, owing to porosity and the presence of frogs in the bricks. Generally, for high-integrity fixings, hammer-in sockets are not recommended.

As noted above, there are various methods of attachment of the cladding fixing to the supporting structure, and the combinations are many and varied, depending on the main structural elements. The following examples of fixing design should give a general guide to methods.

Stone cladding to concrete

There are standard cramp shapes which are used in the retention of stone facings. Provided the basic rules governing material availability and sizes are observed, these fixings can be easily specified. Their attachment into the concrete backing material can be by bolt, cast-in dovetail slot, cast-in channel or grouted-in fixings (Fig. 37).

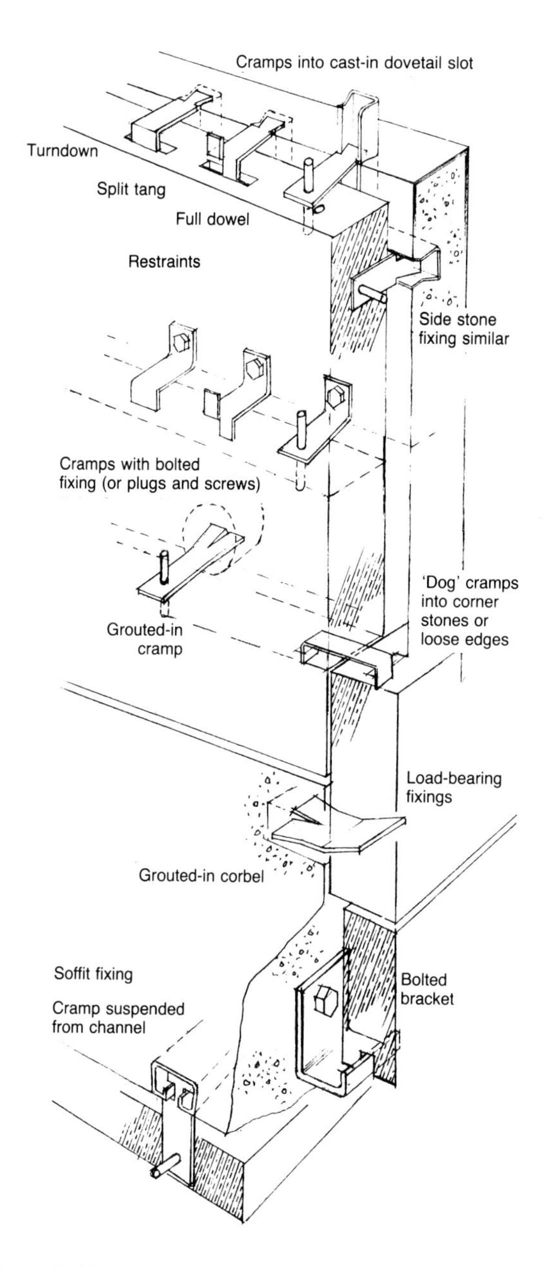
Cramps into cast-in dovetail slot
Turndown
Split tang
Full dowel
Restraints
Side stone fixing similar
Cramps with bolted fixing (or plugs and screws)
Grouted-in cramp
'Dog' cramps into corner stones or loose edges
Load-bearing fixings
Grouted-in corbel
Soffit fixing
Cramp suspended from channel
Bolted bracket

Fig. 37. Stone cladding to concrete

The dovetail slot should only be used in a cast-in situation and should not be used for load-bearing fixings, nor should it be planted onto a concrete face. Heavy-duty channel fixings can be cast-in, and the loads carried may be determined from the manufacturer's technical data or established through test.

The choice of the type of cramp used is optional from the design point of view, provided any loading onto a dowel pin or turned-over end is acceptable by design. The dowel is generally favoured by stonemasons, as it requires only the use of a drill. A mortice for a turned-over end is more expensive to form, taking longer to cut. In heavy classical-style masonry, great care needs to be taken in detailing, as each detail is likely to vary owing to the form and size of the stones. The use of concrete as a main structural element has made the fixing of traditional stonework slightly more straightforward. However, when fixing to steel-framed structures the methods are still, of necessity, very similar to those of nearly a century ago, and as ideally illustrated in *A modern practical masonry* by E. G. Warland, first published in 1929 and recently republished by the Stone Federation.

In the design of heavy stone soffits, there are governing minimum measurements of stone laid down, and these should be strictly adhered to.

Brick cladding to concrete: brick and brickwork

Brickwork may be either a single-skin facing tied back to concrete or two skins with a cavity between; often the cavity is filled with insulation material (Fig. 38). In either case the wall must be stable and tied with restraints or wall ties at the stipulated centres laid down in the British Standard 5628.

With the smaller individual units, the positioning of wall ties or restraint fixings becomes relatively straightforward. However, brickwork needs to be supported at heights not exceeding every 9 m or three storeys. Ideally, the method of support should be a continuous long angle. There are proprietary individual systems on the market, but care needs to be taken when using the single-brick system to ensure that the necessary fire protection is afforded over openings.

Any fixing passing across a cavity is normally required to contain a drip or moisture-shedding element.

When using long angles, the elements should be manufactured

in practical lengths and there should be a gap between each angle of about 10 mm. Curved angles may be manufactured, but the radius should be kept at a reasonable dimension to avoid distortion. With stainless steel angles which are formed, i.e. bent in one piece, the ability to roll the section to a curve on the opposing plane is extremely limited, and distortion and locked-in stress in the material becomes a problem.

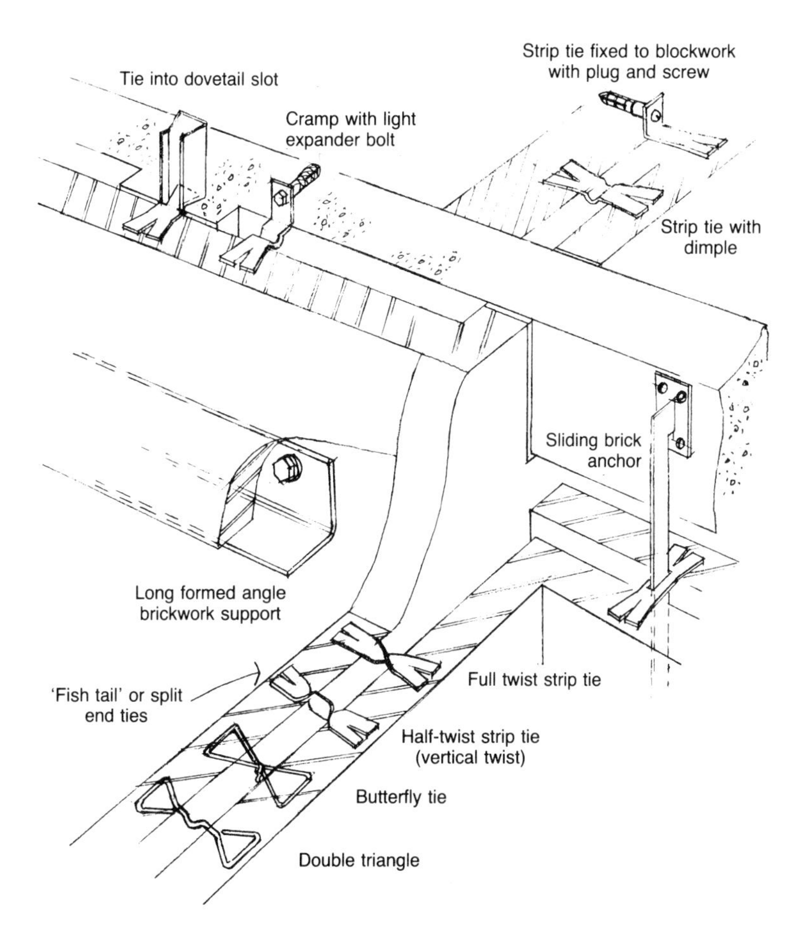

Fig. 38. Brick facing to concrete brick and block. It should be noted that double-ended 'fish tails' (split and ties) are traditional, but modern safety ties of varying design are generally recommended, especially where ties are to be built into one leaf in advance of the second leaf

In designing angles to wrap around columns or corners, one must try to avoid contained dimensions, as site tolerance may be a problem in fixing. Mitred corners may be used, or straight-welded plates. One should check deflections on all extra length corner plates to avoid cracking of brickwork. Gussets may be used to stiffen difficult brackets, but it is important to avoid too much cutting of bricks.

Stone to brick or blockwork: cast stone

The fixings are very similar in principal in all these cases, except that care must be exercised if loads from supported stonework are to be carried by brickwork. The wall must be thick enough

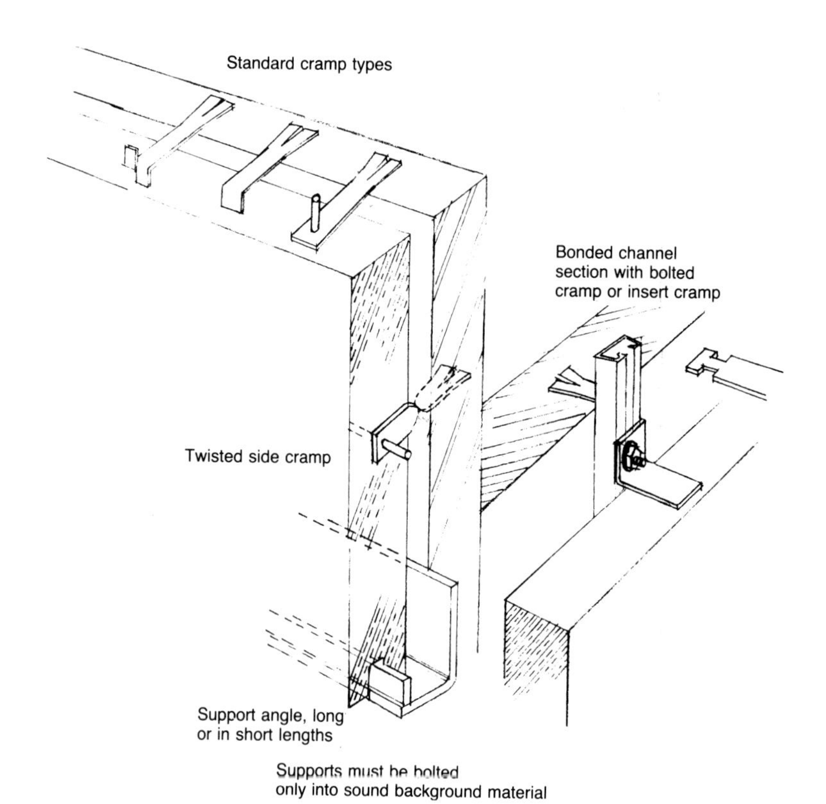

Fig. 39. Stonework/brick and blockwork

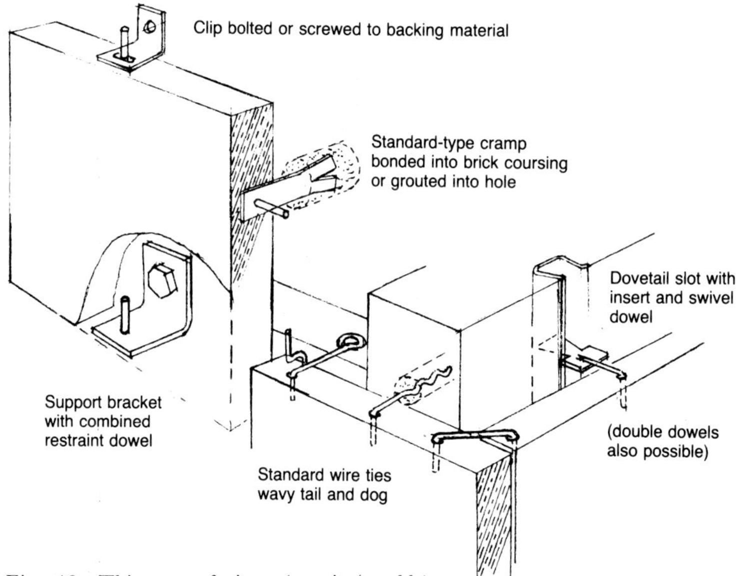

Fig. 40. Thin stone facings (granite/marble)

to take the eccentricity induced by loading from a supported facing panel (Fig. 39).

Cast stone is normally fixed as natural stone, but as this type of cladding unit is often reinforced it may be treated similarly to precast concrete, and can be hung from beams or parapets instead of being bottom-supported.

Thin facings (generally internal)

These are treated in a very similar manner to natural heavy stone masonry, except that the fixings are generally of lighter section and may be made with wire or dowels. As in all supported elements, compression joints must be provided under the support (Fig. 40).

Precast concrete

This field tends to be specialized and individual, and each type of unit needs to be considered on its own merits. However, basic principles should be observed, and more especially so because of the heavy loads imposed by precast units. Support should only be on two points unless the unit is sitting directly onto a continuous

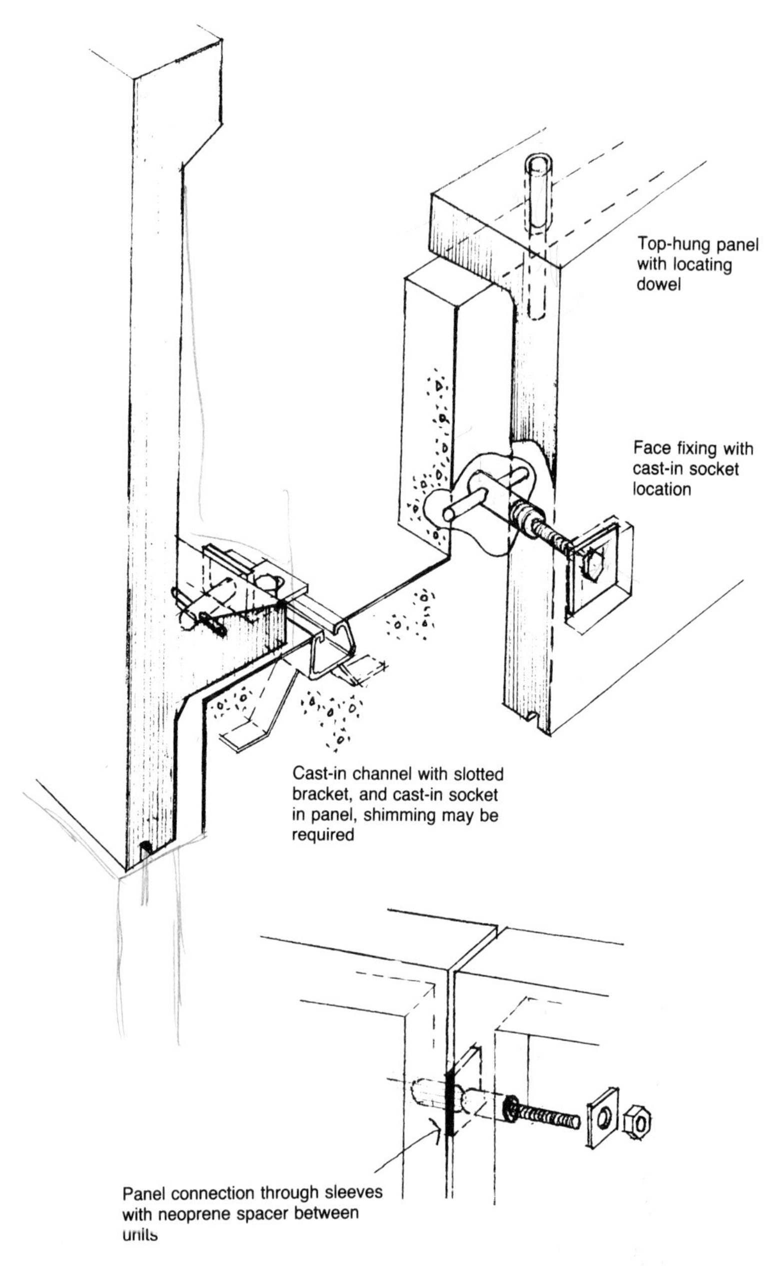
Top-hung panel
with locating
dowel
Face fixing with
cast-in socket
location
Cast-in channel with slotted
bracket, and cast-in socket
in panel, shimming may be
required
Panel connection through sleeves
with neoprene spacer between
units

Fig. 41. Precast panels

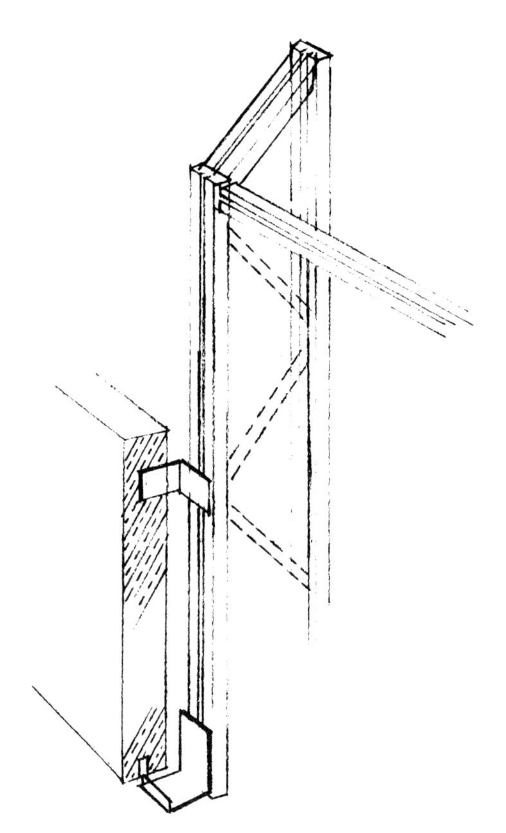

Fig. 42. Framing systems

floor edge, or something similar. Even then, level bedding of the unit must be ensured. Cast-in fixings must be well tied into the unit and should penetrate beyond any reinforcement; they should not be placed at or near an edge (Fig. 41).

Account must be taken of thermal movement in precast units. Thermal stress, when calculated, will be found to be enormously high, and movement can have a devastating effect on the unit and fixing alike.

Framing systems

It is often necessary to support facing panels at some distance from the main structure, generally to provide architectural features,

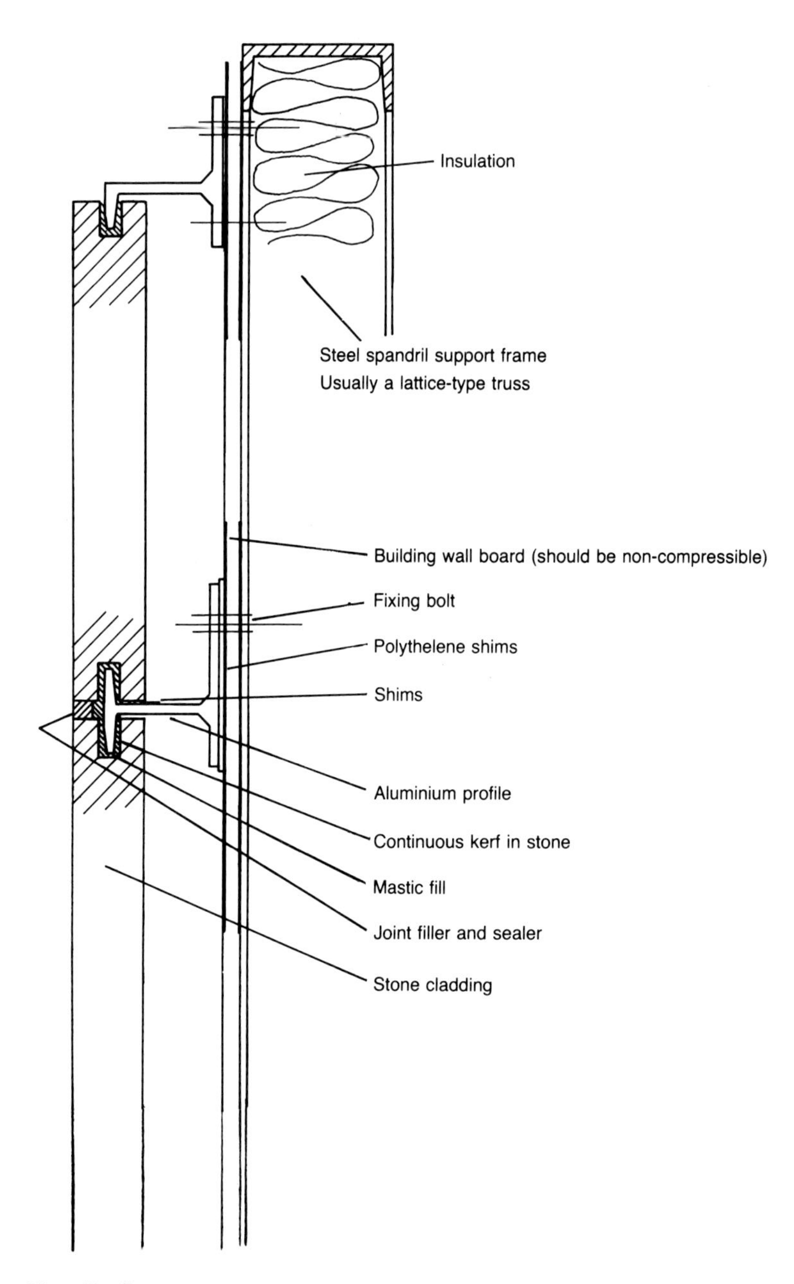

Fig. 43. Stone truss system

when it is too expensive or impractical to carry the backing structure out to that line. In these cases structural frame may be made up to suit the design (Fig. 42). These become structural elements in themselves, and, depending on the loads considered, may be fabricated from either standard channel sections or structural steelwork.

The advantage of using channels is that they may be obtained in stainless steel, whereas stainless steel structural sections are expensive and difficult to obtain. Whenever using mild steel elements with stainless steel fixings, isolators must be used to avoid bimetallic corrosion.

GRC

The fixing requirements for GRC are very similar to those for precast concrete, except that loads will generally be lighter. However, movement is again an important consideration, and deflections are generally much larger on GRC units, inducing high stress and dimensional movement on fixings.

In casting fixings into GRC panels a solid block of material must be provided against and around the fixing, especially in sandwich panel construction. GRC, like precast concrete, is weak in tension and should be bottom-supported. As noted above, the long-term strength of GRC must be considered when designing these elements.

Stone truss systems

The design concepts illustrated in Fig. 43 are based on traditional methods employed in the UK. They are dependent on a backing material, masonry or concrete, which forms the inner skin of the whole environmental overcoat. This inner skin is supported on structural members spanning between columns or vertical supports; the loads are therefore heavy, and together with the weight of cladding it may be seen that loads are, in some instances, supported twice.

The modern design philosophy in the USA and parts of Europe, especially Scandinavia, is vastly different. It is important to understand this alternative system, as projects are now being constructed in the UK using such methods.

The initial concept is to cut the thickness of cladding stonework to the almost dangerous minimum; in some cases thicknesses come

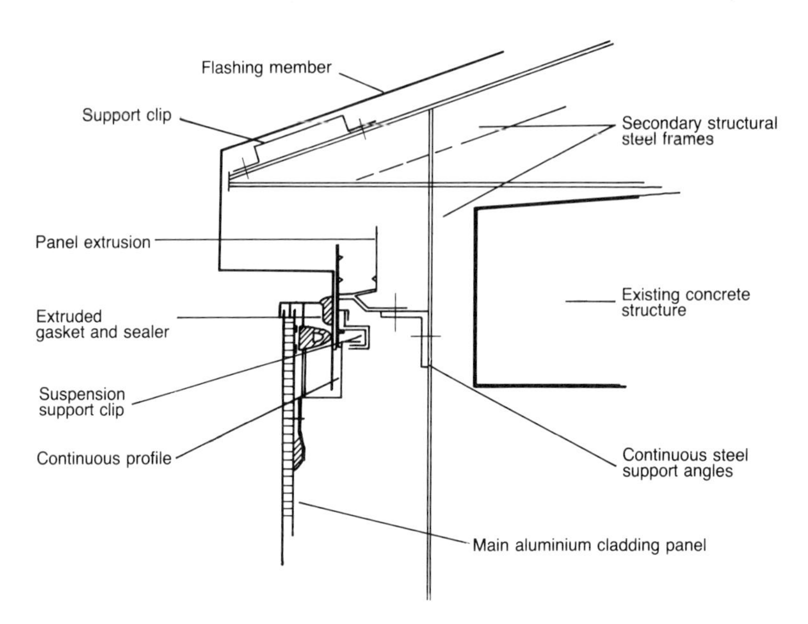

Fig. 44. Patent metal cladding

below those recommended by British Standards. However, with advent of Eurocodes such requirements could well be varied in future. Once the stonework has been reduced to such minimal dimensions, the use of traditional fixings is not quite so simple or even advisable. The approach is therefore to cut continuous grooves (or kerfs) in the stones, and to use lengths of angle retainers fixed back to a secondary structure, usually steelwork. The stone retention fixings are often aluminium.

There are basic design concepts which are at variance with traditional methods. The stone is actually designed by calculation using figures, based on tests, for flexual and shear strength. Such methods can give misleadingly low design stresses even when using the concept of high factors of safety applied to the working stresses.

Two very important factors need to be considered: cutting and manufacturing tolerances, and packing and transportation methods. There is a high incidence of failure due to damage of the thin clutch material on the edges of the stone. If the strengths of the stones can be justified by calculation to accept applied wind-imposed or other lateral loads, the next design problem is the backing structure.

The philosophy is to mount the stonework on structural frames which span directly onto the columns of the main structure. These frames are then insulated and faced internally to provide an environmental and insulated inner panel. There are therefore no edge beams in the main structure, nor internal walls to be supported.

The design therefore lightens loads on the building and thus on the foundations. For very tall buildings significant savings may be justified. The structural design concept is also one suited to earthquake-resisting structural design.

One of the main fixing problems in such designs is the attachment of the stone support angles, which are generally aluminium, back to the steel subframes. It is common practice to use self-tapping screws into the steelwork, and sometimes also passing through a facing board over the steelwork. This, by UK standards, is not good or acceptable practice. Self-tapping screws may become loosened under relatively light applied loads, and they are not generally recommended for load-bearing situations; therefore when considering this alternative stone-fixing method great attention must be given to the method of fixing.

Another consideration is that of the deflection of the frames. The steel subframe may be very long, and a deflection of even a small magnitude could cause crushing stresses on the stone panels. Remembering that the stones are thin and that the clutch material is slender, stone failures can and do occur. There is a further complication, in that the frames should be free to move on at least one support; if they cannot, thermal stresses may also be imparted to the stone.

It is often a practice to prefix the stone to the steel frames and to hoist the entire unit into position. Such methods dictate the stiffness of such frames if damage is to be avoided.

It is also important to consider the stone quality. Thin stones, about 20–30 mm, may move considerably. Granite is relatively stable, but some marbles move considerably and flex and bend, inducing unwanted stresses in the stones. There have been some dramatic failures, and replacement of the stones is very difficult when there are long continuous angles in the kerfs. When investigating problems allied to this method of design, note should be taken of the likely difficulties involved. When designing a new façade using the method described, care must be taken to ensure that all the above points are taken into consideration.

Metal cladding

The refurbishment of existing buildings is often achieved by means of overcladding. Such systems generally comprise lightweight metal panels, usually in decorative aluminium. The fixing of the new materials is accomplished by attaching secondary framing to the existing structure and fixing the cladding material to the framing. The important consideration in this system of overcladding is the integrity of the secondary framing and its attachment to the existing structure. The suitability of the existing structural members to accept additional loads and fixings is of prime importance.

The methods of cladding present several major design factors which will need to be considered, not only at the initial erection stage but also in the event of 'failure' investigation.

Movement between the new and existing structures must be compatible, and, although the jointing in the new metal facing will be considered separately in order to suit the cladding elements, the supporting framework will need to reflect the main building joints. The secondary framework will need to be stiff enough to avoid excessive deflection of the new cladding.

The next important aspect is the attachment of dissimilar metals, one to another. A major factor in such designs, and one often overlooked, is that of tolerance. Factory-made products such as standard panels and cleat fixings provide accurate outer facing. However, an existing building will have constructional inaccuracies, and the subframes need to be sufficiently flexible to take up such building tolerances, and at the same time effectively match the factory-made product.

A very accurate measurement survey and thorough material investigation of the existing building is absolutely essential. The biggest problem in the new cladding is to provide a watertight shield to avoid deterioration of the existing structure, which will not be accessible once the building has been reclad. Ventilation and insulation are important aspects when considering such refurbishment situations.

When investigating this type of clad structure, these design aspects must be understood before any assessment of the reason for a failure can be made.

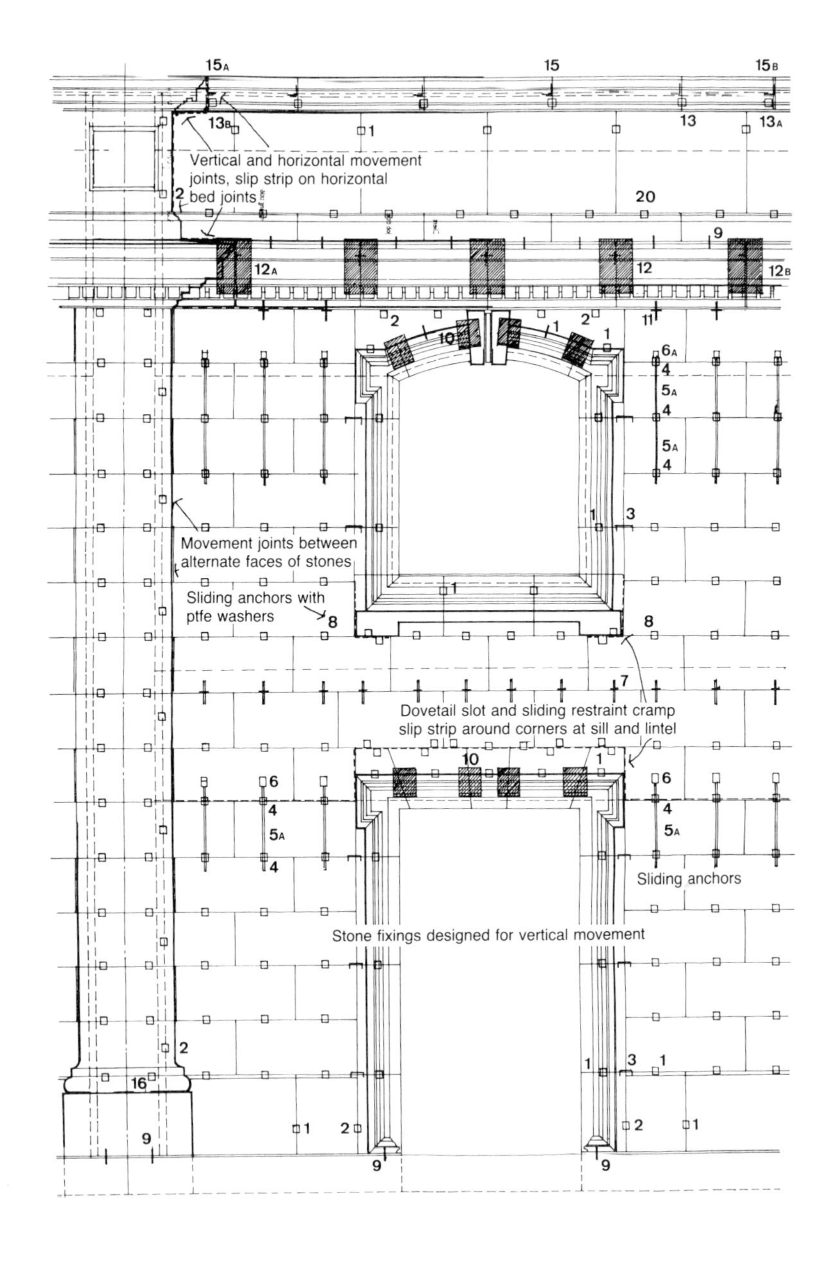
15A
15
15B
13B
1
13
13A
Vertical and horizontal movement joints, slip strip on horizontal bed joints
2
20
9
12A
12
12B
2
10
1
2
1
11
6A
4
5A
4
5A
4
1
3
Movement joints between alternate faces of stones
1
Sliding anchors with ptfe washers
8
8
7
Dovetail slot and sliding restraint cramp slip strip around corners at sill and lintel
10
1
6
4
5A
4
6
4
5A
Sliding anchors
Stone fixings designed for vertical movement
2
1
3
1
16
1
2
2
1
9
9
9

Appendix 1

The illustration shows a typical classical style stone elevation. The fixings have been designed to allow vertical movement, and the corners of windows and sills have horizontal sliding surfaces. Stones recessed behind pilasters also have movement joints provided at alternate faces to suit bonding.

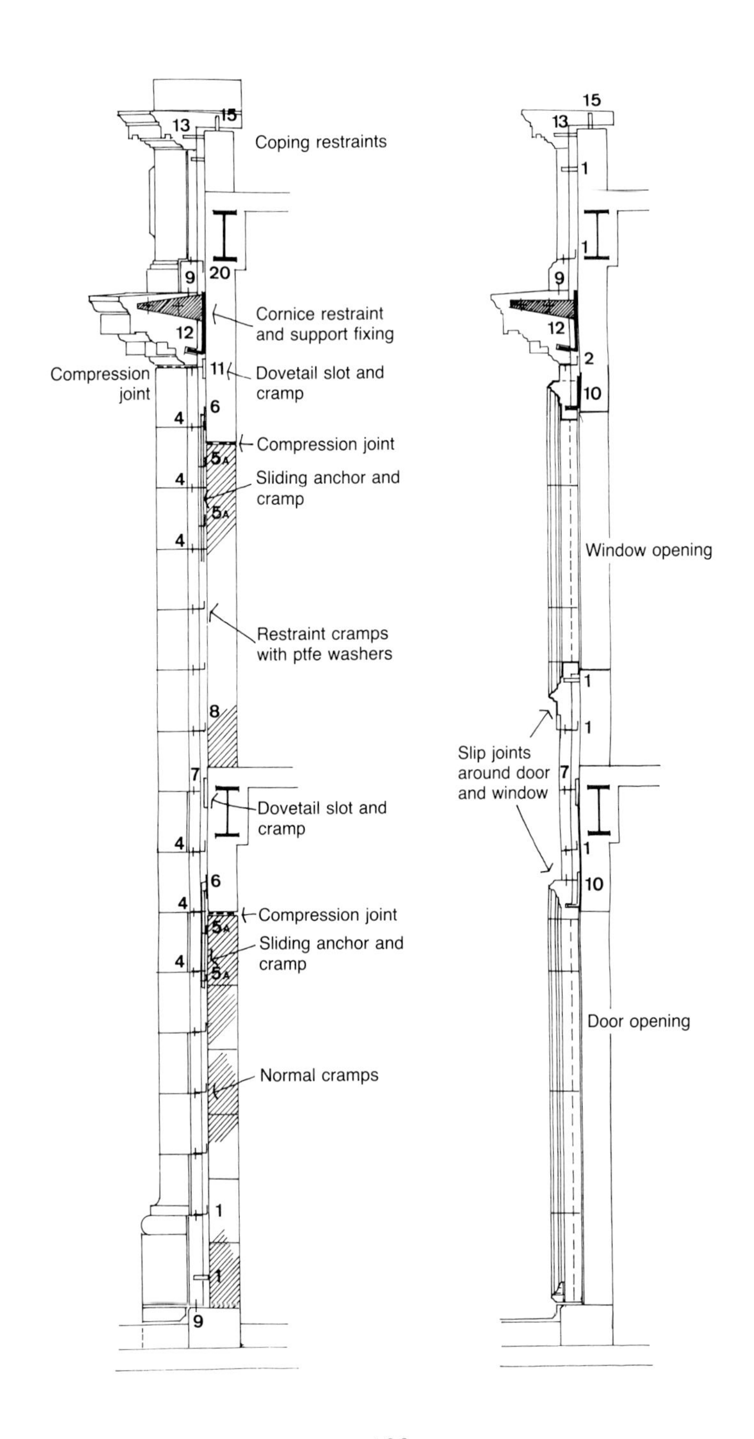

15
13
Coping restraints
20
9
Cornice restraint
and support fixing
12
Compression
joint
11
Dovetail slot and
cramp
6
4
Compression joint
5A
4
Sliding anchor and
cramp
5A
4
Restraint cramps
with ptfe washers
8
7
Dovetail slot and
cramp
4
6
4
Compression joint
5A
4
Sliding anchor and
cramp
5A
Normal cramps
1
1
9
15
13
1
1
9
12
2
10
Window opening
1
1
Slip joints
around door
and window
7
1
10
Door opening

Appendix 2

The illustrations indicate typical sectional detailing through the main elevation.

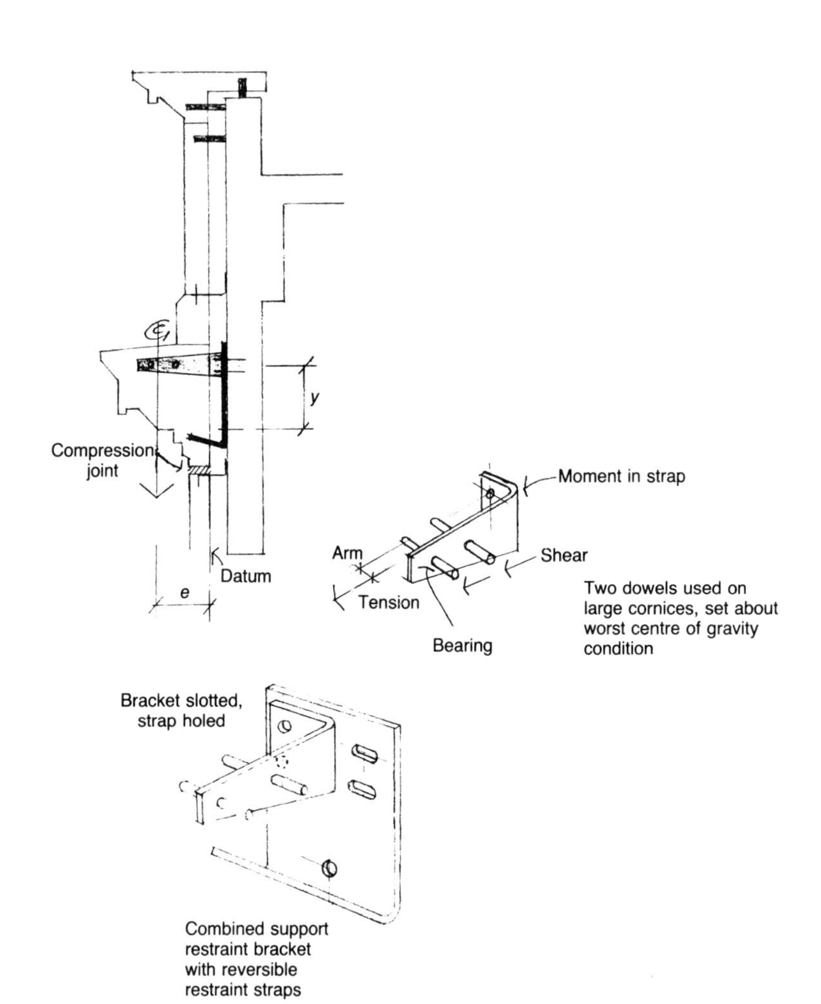
y
Compression joint
Datum
e
Moment in strap
Arm
Shear
Tension
Bearing
Two dowels used on large cornices, set about worst centre of gravity condition
Bracket slotted, strap holed
Combined support restraint bracket with reversible restraint straps

Appendix 3

Described below are the design considerations for a large projecting cornice stone. It is important to ensure that the restraint strap adequately ties the stone about the centre of gravity. The load must not cause excessive deflection in the load-bearing bracket.

1. Calculate the stone weight and the centre of gravity of mass from a datum.
2. Calculate the full overturning moment using the full dimension from the supporting structure (i.e. e + cavity dimension).
3. The couple is equal to the full overturning moment divided by dimension y.
4. The design moment for the cantilever bracket is the stone load multiplied by the length of the cantilever.
5. The strap must resist the tension force from the couple which induces a moment in the strap.
6. The dowel must resist the shear force from the tension load. One must also check the bearing stress on the strap from the dowels.
7. The supporting bolts at the top carry the tension load as a pull-out force. The bottom bolt is in compression and is a location bolt. All bolts need to take the full vertical load as a shear force.

On large cornice projections it is usual to take a snow allowance or access point load in calculating load conditions.

Appendix 4

The illustrations that follow are load flow diagrams for a typical cladding elevation. It is important to understand the loading conditions to be able to determine the possible causes of damage when inspecting problems.

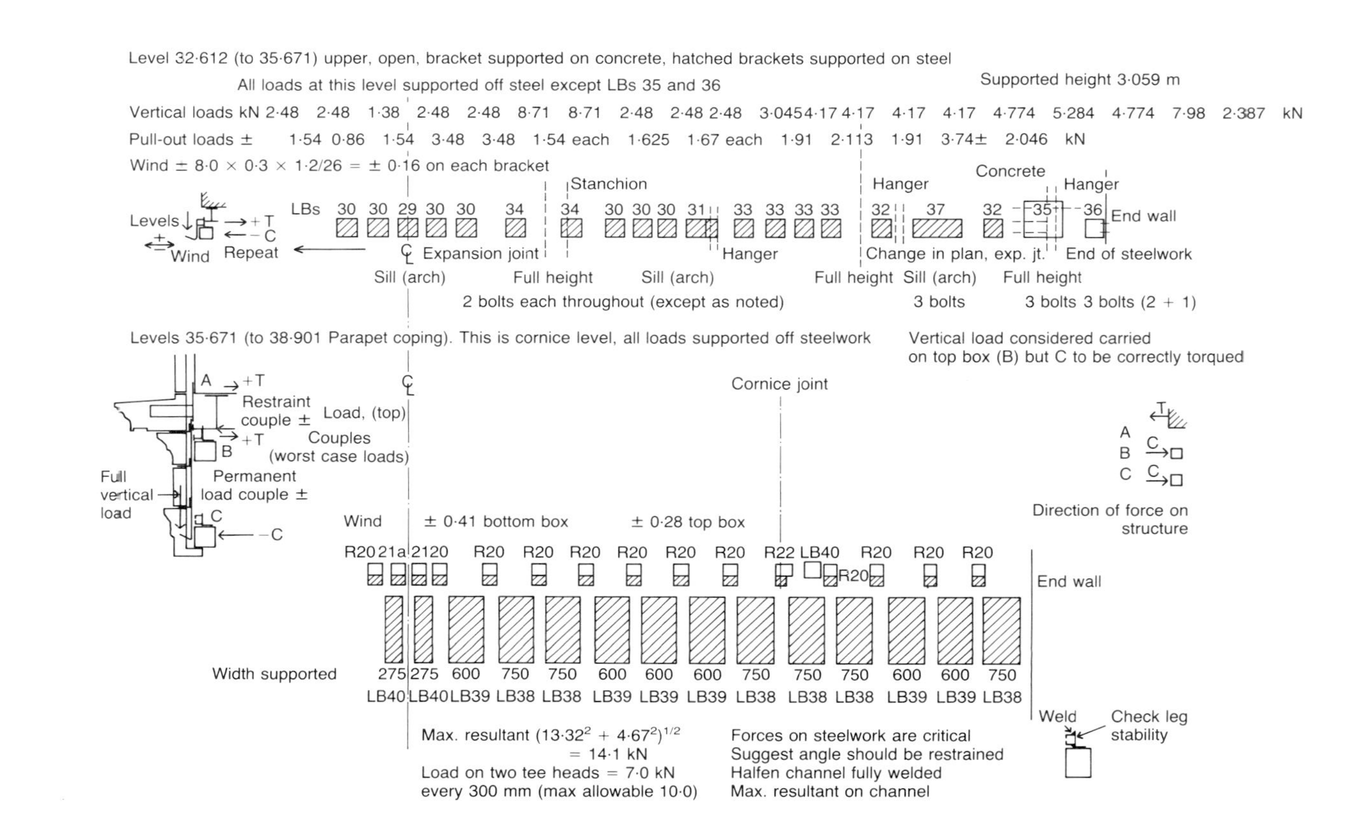
Level 32·612 (to 35·671) upper, open, bracket supported on concrete, hatched brackets supported on steel
All loads at this level supported off steel except LBs 35 and 36
Supported height 3·059 m
Vertical loads kN 2·48 2·48 1·38 2·48 2·48 8·71 8·71 2·48 2·48 2·48 3·045 4·17 4·17 4·17 4·17 4·774 5·284 4·774 7·98 2·387 kN
Pull-out loads ± 1·54 0·86 1·54 3·48 3·48 1·54 each 1·625 1·67 each 1·91 2·113 1·91 3·74± 2·046 kN
Wind ± 8·0 × 0·3 × 1·2/26 = ± 0·16 on each bracket
Levels
Wind
+T
−C
Repeat
LBs 30 30 29 30 30 34 34 30 30 30 31 33 33 33 33 32 37 32 35 36
Stanchion
Hanger
Concrete
Hanger
End wall
Expansion joint
Hanger
Change in plan, exp. jt.
End of steelwork
Sill (arch)
Full height
Sill (arch)
Full height
Sill (arch)
Full height
2 bolts each throughout (except as noted)
3 bolts
3 bolts 3 bolts (2 + 1)
Levels 35·671 (to 38·901 Parapet coping). This is cornice level, all loads supported off steelwork
Vertical load considered carried on top box (B) but C to be correctly torqued
A
+T
Restraint couple ±
Load, (top)
+T
B
Couples (worst case loads)
Full vertical load
Permanent load couple ±
C
−C
Cornice joint
A
B
C
T
C
C
Direction of force on structure
Wind
± 0·41 bottom box
± 0·28 top box
R20 21a 21 20 R20 R20 R20 R20 R20 R20 R22 LB40 R20 R20 R20 R20
End wall
Width supported
275 275 600 750 750 600 600 600 750 750 750 600 600 750
LB40 LB40 LB39 LB38 LB38 LB39 LB39 LB39 LB38 LB38 LB38 LB39 LB39 LB38
Max. resultant $(13{\cdot}32^2 + 4{\cdot}67^2)^{1/2}$
= 14·1 kN
Load on two tee heads = 7·0 kN
every 300 mm (max allowable 10·0)
Forces on steelwork are critical
Suggest angle should be restrained
Halfen channel fully welded
Max. resultant on channel
Weld
Check leg stability

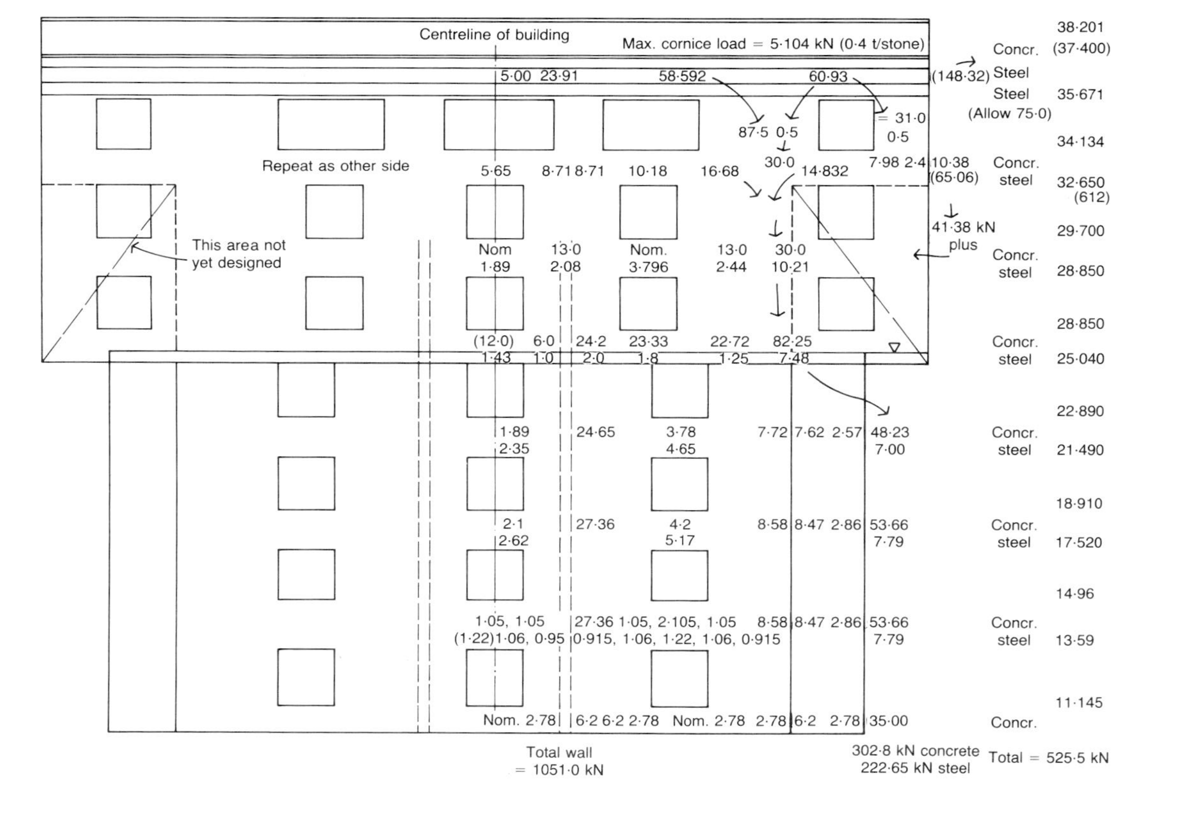

Centreline of building
Max. cornice load = 5·104 kN (0·4 t/stone)
Repeat as other side
This area not yet designed
41·38 kN plus
(Allow 75·0)
Total wall = 1051·0 kN
302·8 kN concrete
222·65 kN steel
Total = 525·5 kN

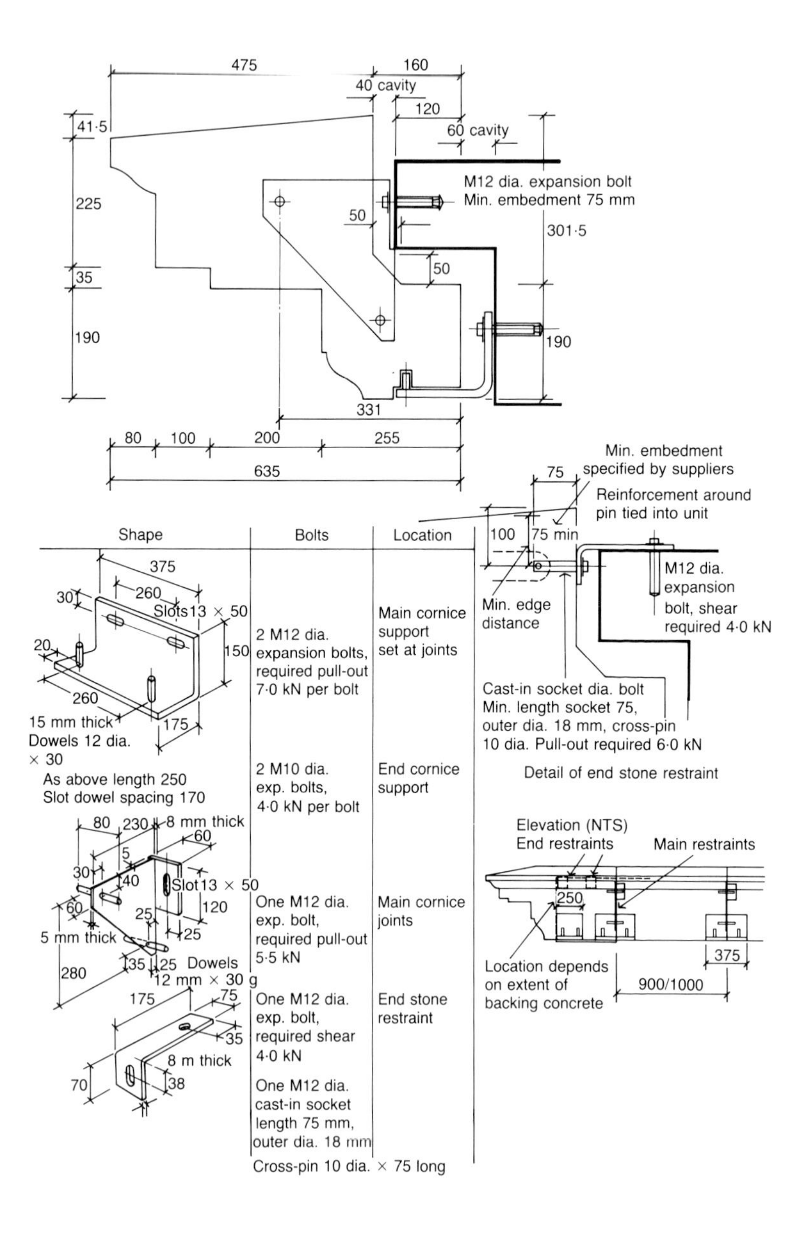

475
160
40 cavity
120
60 cavity
41·5
225
35
190
M12 dia. expansion bolt
Min. embedment 75 mm
50
301·5
50
190
331
80
100
200
255
635
Shape
Bolts
Location
375
260
30
Slots 13 × 50
20
150
260
15 mm thick
175
Dowels 12 dia.
× 30
2 M12 dia.
expansion bolts,
required pull-out
7·0 kN per bolt
Main cornice
support
set at joints
As above length 250
Slot dowel spacing 170
2 M10 dia.
exp. bolts,
4·0 kN per bolt
End cornice
support
80
230
8 mm thick
60
5
30
40
Slot 13 × 50
60
120
25
5 mm thick
25
280
35
25
Dowels
12 mm × 30 g
One M12 dia.
exp. bolt,
required pull-out
5·5 kN
Main cornice
joints
175
75
35
8 m thick
70
38
One M12 dia.
exp. bolt,
required shear
4·0 kN
End stone
restraint
One M12 dia.
cast-in socket
length 75 mm,
outer dia. 18 mm
Cross-pin 10 dia. × 75 long
Min. embedment
specified by suppliers
75
Reinforcement around
pin tied into unit
100
75 min
Min. edge
distance
M12 dia.
expansion
bolt, shear
required 4·0 kN
Cast-in socket dia. bolt
Min. length socket 75,
outer dia. 18 mm, cross-pin
10 dia. Pull-out required 6·0 kN
Detail of end stone restraint
Elevation (NTS)
End restraints
Main restraints
250
Location depends
on extent of
backing concrete
375
900/1000

Appendix 5

The illustration represents a cast stone element used in a cornice situation. In this case the restraint gusset in the vertical joint restrains both sections of the stone across the narrow neck necessitated by the design requirements. Attention to details of this nature is necessary to avoid future problems. Fracture of the unit across the neck would cause the upper part to become unstable.

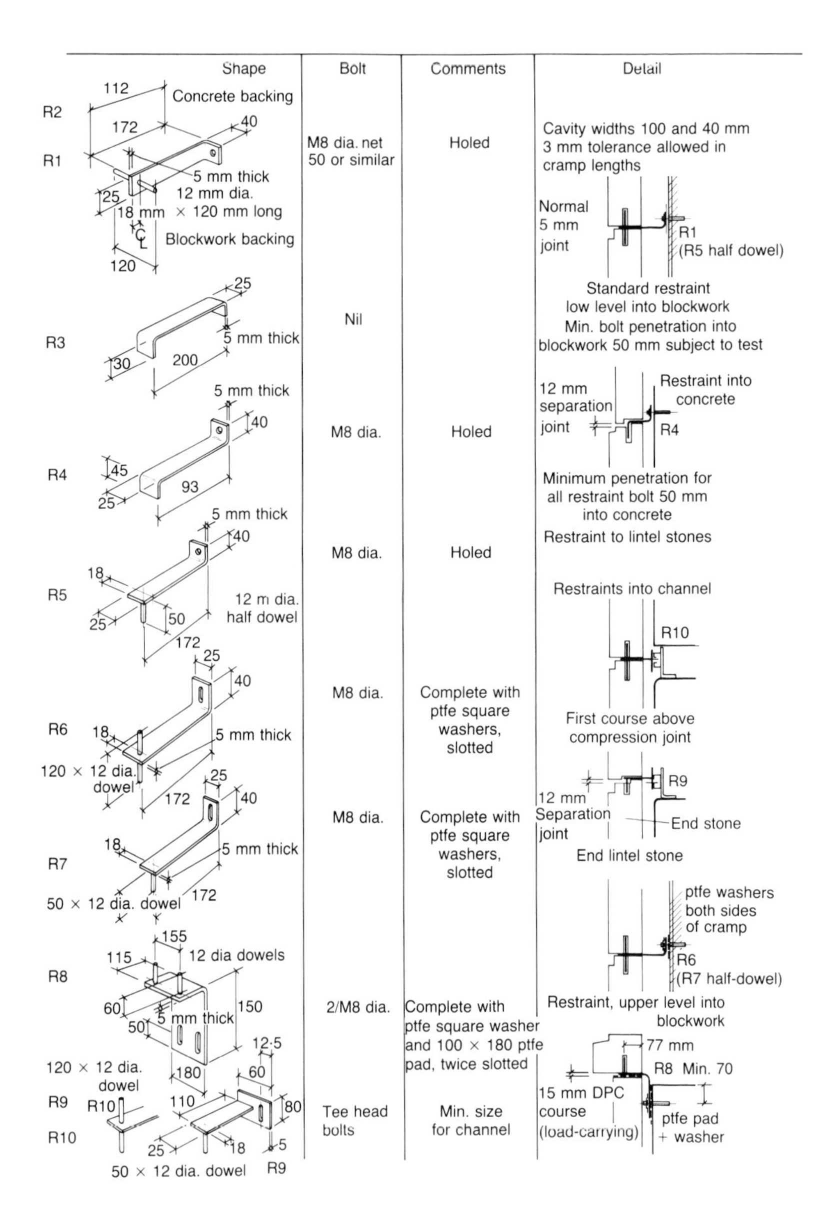

	Shape	Bolt	Comments	Detail
R2 R1	112; Concrete backing; 172; 40; 5 mm thick; 12 mm dia. × 120 mm long; 25; 18 mm; ℄; Blockwork backing; 120	M8 dia. net 50 or similar	Holed	Cavity widths 100 and 40 mm 3 mm tolerance allowed in cramp lengths Normal 5 mm joint; R1 (R5 half dowel) Standard restraint low level into blockwork
R3	25; 5 mm thick; 30; 200	Nil		Min. bolt penetration into blockwork 50 mm subject to test
R4	5 mm thick; 40; 45; 93; 25	M8 dia.	Holed	12 mm separation joint; Restraint into concrete; R4 Minimum penetration for all restraint bolt 50 mm into concrete
R5	5 mm thick; 40; 18; 12 m dia. half dowel; 25; 50; 172	M8 dia.	Holed	Restraint to lintel stones
R6	25; 40; 18; 5 mm thick; 120 × 12 dia. dowel; 172	M8 dia.	Complete with ptfe square washers, slotted	Restraints into channel; R10 First course above compression joint
R7	25; 40; 18; 5 mm thick; 50 × 12 dia. dowel; 172	M8 dia.	Complete with ptfe square washers, slotted	R9; 12 mm Separation joint; End stone End lintel stone ptfe washers both sides of cramp; R6 (R7 half-dowel)
R8	155; 115; 12 dia dowels; 60; 5 mm thick; 150; 50; 12·5; 120 × 12 dia. dowel; 180; 60	2/M8 dia.	Complete with ptfe square washer and 100 × 180 ptfe pad, twice slotted	Restraint, upper level into blockwork 77 mm; R8 Min. 70; 15 mm DPC course (load-carrying); ptfe pad + washer
R9 R10	R10; 110; 80; 25; 18; 5; 50 × 12 dia. dowel; R9	Tee head bolts	Min. size for channel	

Appendix 6

The details illustrate a comprehensive method of detailing which is suitable for all stages of costing, manufacture and site management control. Full construction details similar to these are essential for main construction purposes.

References

1. British Standards Institution. *Use of masonry*. BSI, London, 1978. BS 5628.
2. British Standards Institution. *Design and installation of natural stone cladding and lining*. BSI, London, 1989. BS 8298.
3. Stone Federation. *Code of practice on natural stone cladding (non-load bearing)*. London, 1984.
4. British Standards Institution. *Code of practice for stone masonry*. BSI, London, 1976. BS 5390.
5. British Standards Institution. *Reconstructed stone masonry units*. BSI, London, 1984. BS 6457.
6. British Standards Institution. *Cast stone*. BSI, London, 1986. BS 1217.

The authors would like to thank Harris & Edgar Limited.

Index